A COURSE IN ALGEBRAIC NUMBER THEORY

ROBERT B. ASH

Department of Mathematics
University of Illinois

DOVER PUBLICATIONS, INC.
Mineola, New York

Bibliographical Note

This Dover edition, first published in 2010, is the first publication in book form of *A Course in Algebraic Number Theory,* which is available from the author's website at http://www.math.uiuc.edu/~r-ash/ANT.html.

International Standard Book Number

ISBN-13: 978-0-486-47754-1
ISBN-10: 0-486-47754-1

Manufactured in the United States by Courier Corporation
47754101
www.doverpublications.com

Preface

This is a text for a basic course in algebraic number theory, written in accordance with the following objectives.

1. Provide reasonable coverage for a one-semester course.

2. Assume as prerequisite a standard graduate course in algebra, but cover integral extensions and localization before beginning algebraic number theory. For general algebraic background, see my online text "Abstract Algebra: The Basic Graduate Year", which can be downloaded from my web site www.math.uiuc.edu/~r-ash/*. The abstract algebra material is referred to in this text as TBGY.

3. Cover the general theory of factorization of ideals in Dedekind domains, as well as the number field case.

4. Do some detailed calculations illustrating the use of Kummer's theorem on lifting of prime ideals in extension fields.

5. Give enough details so that the reader can navigate through the intricate proofs of the Dirichlet unit theorem and the Minkowski bounds on element and ideal norms.

6. Cover the factorization of prime ideals in Galois extensions.

7. Cover local as well as global fields, including the Artin-Whaples approximation theorem and Hensel's lemma.

Especially helpful to me in preparing this work were the beautiful little book by Samuel, "Algebraic Theory of Numbers", Hermann 1971, and the treatment of cyclotomic fields by J. Milne in his online text "Algebraic Number Theory" (www.math.lsa.umich.edu/~jmilne/) Some other useful references are:

Esmonde, J., and Murty, M.R., "Problems in Algebraic Number Theory", Springer 1999

Frölich, A., and Taylor, M.J., "Algebraic Number Theory", Cambridge 1991

Janusz, G.J.," Algebraic Number Fields", AMS 1996

Koch, H., "Number Theory", AMS 2000

Marcus, D.A., "Number Fields", Springer 1977

Stewart, I., and Tall, D., "Algebraic Number Theory", Chapman and Hall 1987

*This text is also available from Dover Publications as Basic Abstract Algebra (0-486-45356-1).

Table of Contents

Chapter 1

Introduction

Techniques of abstract algebra have been applied to problems in number theory for a long time, notably in the effort to prove Fermat's last theorem. As an introductory example, we will sketch a problem for which an algebraic approach works very well. If p is an odd prime and $p \equiv 1 \bmod 4$, we will prove that p is the sum of two squares, that is, p can expressed as $x^2 + y^2$ where x and y are integers. Since $\frac{p-1}{2}$ is even, it follows that -1 is a quadratic residue (that is, a square) mod p. To see this, pair each of the numbers $2, 3, \ldots, p-2$ with its inverse mod p, and pair 1 with $p - 1 \equiv -1 \bmod p$. The product of the numbers 1 through $p - 1$ is, mod p,

$$1 \times 2 \times \cdots \times \frac{p-1}{2} \times -1 \times -2 \times \cdots \times -\frac{p-1}{2}$$

and therefore

$$[(\frac{p-1}{2})!]^2 \equiv -1 \mod p.$$

If $-1 \equiv x^2 \bmod p$, then p divides $x^2 + 1$. Now we enter the ring $\mathbb{Z}[i]$ of Gaussian integers and factor $x^2 + 1$ as $(x + i)(x - i)$. Since p can divide neither factor, it follows that p is not prime in $\mathbb{Z}[i]$. Since the Gaussian integers form a unique factorization domain, p is not irreducible, and we can write $p = \alpha\beta$ where neither α nor β is a unit.

Define the *norm* of $\gamma = a + bi$ as $N(\gamma) = a^2 + b^2$. Then $N(\gamma) = 1$ iff γ is 1,-1,i or $-i$, equivalently, iff γ is a unit. Thus

$$p^2 = N(p) = N(\alpha)N(\beta) \text{ with } N(\alpha) > 1 \text{ and } N(\beta) > 1,$$

so $N(\alpha) = N(\beta) = p$. If $\alpha = x + iy$, then $p = x^2 + y^2$.

Conversely, if p is an odd prime and $p = x^2 + y^2$, then p is congruent to 1 mod 4. [If x is even, then $x^2 \equiv 0 \bmod 4$, and if x is odd, then $x^2 \equiv 1 \bmod 4$. We cannot have x and y both even or both odd, since p is odd.]

It is natural to conjecture that we can identify those primes that can be represented as $x^2 + |m|y^2$, where m is a negative integer, by working in the ring $\mathbb{Z}[\sqrt{m}]$. But the above argument depends critically on unique factorization, which does not hold in general. A standard example is $2 \times 3 = (1 + \sqrt{-5})(1 - \sqrt{-5})$ in $\mathbb{Z}[\sqrt{-5}]$. Difficulties of this sort led Kummer to invent "ideal numbers", which became ideals at the hands of Dedekind. We will see that although a ring of algebraic integers need not be a UFD, unique factorization of ideals will always hold.

1.1   Integral Extensions

If E/F is a field extension and $\alpha \in E$, then α is algebraic over F iff α is a root of a nonconstant polynomial with coefficients in F. We can assume if we like that the polynomial is monic, and this turns out to be crucial in generalizing the idea to ring extensions.

1.1.1   Definitions and Comments

All rings are assumed commutative. Let A be a subring of the ring R, and let $x \in R$. We say that x is *integral over* A if x is a root of a monic polynomial f with coefficients in A. The equation $f(X) = 0$ is called an *equation of integral dependence* for x over A. If x is a real or complex number that is integral over $\mathbb{Z}$, then x is called an *algebraic integer*. Thus for every integer d, $\sqrt{d}$ is an algebraic integer, as is any n^{th} root of unity. (The monic polynomials are, respectively, $X^2 - d$ and $X^n - 1$.) The next results gives several conditions equivalent to integrality.

1.1.2   Theorem

Let A be a subring of R, and let $x \in R$. The following conditions are equivalent:
(i) The element x is integral over A;
(ii) The A-module $A[x]$ is finitely generated;
(iii) The element x belongs to a subring B of R such that $A \subseteq B$ and B is a finitely generated A-module;
(iv) There is a subring B of R such that B is a finitely generated A-module and x *stabilizes* B, that is, $xB \subseteq B$. (If R is a field, the assumption that B is a subring can be dropped, as long as $B \neq 0$);
(v) There is a faithful $A[x]$-module B that is finitely generated as an A-module. (Recall that a faithful module is one whose annihilator is 0.)

Proof.

(i)implies (ii): If x is a root of a monic polynomial of degree n over A, then x^n and all higher powers of x can be expressed as linear combinations of lower powers of x. Thus $1, x, x^2, \ldots, x^{n-1}$ generate $A[x]$ over A.

(ii) implies (iii): Take $B = A[x]$.

(iii) implies (i): If $\beta_1, \ldots, \beta_n$ generate B over A, then $x\beta_i$ is a linear combination of the β_j, say $x\beta_i = \sum_{j=1}^{n} c_{ij}\beta_j$. Thus if β is a column vector whose components are the β_i, I is an n by n identity matrix, and $C = [c_{ij}]$, then

$$(xI - C)\beta = 0,$$

and if we premultiply by the adjoint matrix of $xI - C$ (as in Cramer's rule), we get

$$[\det(xI - C)]I\beta = 0$$

hence $\det(xI - C)b = 0$ for every $b \in B$. Since B is a ring, we may set $b = 1$ and conclude that x is a root of the monic polynomial $\det(XI - C)$ in $A[X]$.

If we replace (iii) by (iv), the same proofs work. If R is a field, then in (iv)⇒(i), x is an eigenvalue of C, so $\det(xI - C) = 0$.

If we replace (iii) by (v), the proofs go through as before. [Since B is an $A[x]$-module, in (v)⇒(i) we have $x\beta_i \in B$. When we obtain $[\det(xI - C)]b = 0$ for every $b \in B$, the hypothesis that B is faithful yields $\det(xI - C) = 0$.] ♣

We are going to prove a transitivity property for integral extensions, and the following result will be helpful.

1.1.3 Lemma

Let A be a subring of R, with $x_1, \ldots, x_n \in R$. If x_1 is integral over A, x_2 is integral over $A[x_1], \ldots$, and x_n is integral over $A[x_1, \ldots, x_{n-1}]$, then $A[x_1, \ldots, x_n]$ is a finitely generated A-module.

Proof. The $n = 1$ case follows from (1.1.2), condition (ii). Going from $n - 1$ to n amounts to proving that if A, B and C are rings, with C a finitely generated B-module and B a finitely generated A-module, then C is a finitely generated A-module. This follows by a brief computation:

$$C = \sum_{j=1}^{r} By_j, \ B = \sum_{k=1}^{s} Ax_k, \text{ so } C = \sum_{j=1}^{r}\sum_{k=1}^{s} Ay_jx_k. \ ♣$$

1.1.4 Transitivity of Integral Extensions

Let A, B and C be subrings of R. If C is integral over B, that is, every element of C is integral over B, and B is integral over A, then C is integral over A.

Proof. Let $x \in C$, with $x^n + b_{n-1}x^{n-1} + \cdots + b_1x + b_0 = 0$, $b_i \in B$. Then x is integral over $A[b_0, \ldots, b_{n-1}]$. Each b_i is integral over A, hence over $A[b_0, \ldots, b_{i-1}]$. By (1.1.3), $A[b_0, \ldots, b_{n-1}, x]$ is a finitely generated A-module. It follows from condition (iii) of (1.1.2) that x is integral over A. ♣

1.1.5 Definitions and Comments

If A is a subring of R, the *integral closure* of A in R is the set A_c of elements of R that are integral over A. Note that $A \subseteq A_c$ because each $a \in A$ is a root of $X - a$. We say that A is *integrally closed* in R if $A_c = A$. If we simply say that A is *integrally closed* without reference to R, we assume that A is an integral domain with fraction field K, and A is integrally closed in K.

If x and y are integral over A, then just as in the proof of (1.1.4), it follows from (1.1.3) that $A[x, y]$ is a finitely generated A-module. Since $x + y, x - y$ and xy belong to this module, they are integral over A by (1.1.2), condition (iii). The important conclusion is that

$$A_c \text{ is a subring of } R \text{ containing } A.$$

If we take the integral closure of the integral closure, we get nothing new.

1.1.6 Proposition

The integral closure A_c of A in R is integrally closed in R.

Proof. By definition, A_c is integral over A. If x is integral over A_c, then as in the proof of (1.1.4), x is integral over A, and therefore $x \in A_c$. ♣

We can identify a large class of integrally closed rings.

1.1.7 Proposition

If A is a UFD, then A is integrally closed.

Proof. If x belongs to the fraction field K, then we can write $x = a/b$ where $a, b \in A$, with a and b relatively prime. If x is integral over A, then there is an equation of the form

$$(a/b)^n + a_{n-1}(a/b)^{n-1} + \cdots + a_1(a/b) + a_0 = 0$$

with all a_i belonging to A. Multiplying by b^n, we have $a^n + bc = 0$, with $c \in A$. Thus b divides a^n, which cannot happen for relatively prime a and b unless b has no prime factors at all, in other words, b is a unit. But then $x = ab^{-1} \in A$. ♣

Problems For Section 1.1

Let A be a subring of the integral domain B, with B integral over A. In Problems 1-3, we are going to show that A is a field if and only if B is a field.

1. Assume that B is a field, and let a be a nonzero element of A. Then since $a^{-1} \in B$, there is an equation of the form

$$(a^{-1})^n + c_{n-1}(a^{-1})^{n-1} + \cdots + c_1 a^{-1} + c_0 = 0$$

with all c_i belonging to A. Show that $a^{-1} \in A$, proving that A is a field.

2. Now assume that A is a field, and let b be a nonzero element of B. By condition (ii) of (1.1.2), $A[b]$ is a finite-dimensional vector space over A. Let f be the A-linear transformation on this vector space given by multiplication by b, in other words, $f(z) = bz$, $z \in A[b]$. Show that f is injective.

3. Show that f is surjective as well, and conclude that B is a field.

In Problems 4-6, let A be a subring of B, with B integral over A. Let Q be a prime ideal of B and let $P = Q \cap A$.

4. Show that P is a prime ideal of A, and that A/P can be regarded as a subring of B/Q.

5. Show that B/Q is integral over A/P.

6. Show that P is a maximal ideal of A if and only if Q is a maximal ideal of B.

1.2 Localization

Let S be a subset of the ring R, and assume that S is *multiplicative*, in other words, $0 \notin S$, $1 \in S$, and if a and b belong to S, so does ab. In the case of interest to us, S will be the complement of a prime ideal. We would like to divide elements of R by elements of S to form the *localized ring* $S^{-1}R$, also called the *ring of fractions* of R by S. There is no difficulty when R is an integral domain, because in this case all division takes place in the fraction field of R. Although we will not need the general construction for arbitrary rings R, we will give a sketch. For full details, see TBGY, Section 2.8.

1.2.1 Construction of the Localized Ring

If S is a multiplicative subset of the ring R, we define an equivalence relation on $R \times S$ by $(a, b) \sim (c, d)$ iff for some $s \in S$ we have $s(ad - bc) = 0$. If $a \in R$ and $b \in S$, we define the fraction a/b as the equivalence class of (a, b). We make the set of fractions into a ring in a natural way. The sum of a/b and c/d is defined as $(ad + bc)/bd$, and the product of a/b and c/d is defined as ac/bd. The additive identity is $0/1$, which coincides with $0/s$ for every $s \in S$. The additive inverse of a/b is $-(a/b) = (-a)/b$. The multiplicative identity is $1/1$, which coincides with s/s for every $s \in S$. To summarize:

$S^{-1}R$ is a ring. If R is an integral domain, so is $S^{-1}R$. If R is an integral domain and $S = R \setminus \{0\}$, then $S^{-1}R$ is a field, the *fraction field* of R.

There is a natural ring homomorphism $h : R \to S^{-1}R$ given by $h(a) = a/1$. If S has no zero-divisors, then h is a monomorphism, so R can be embedded in $S^{-1}R$. In particular, a ring R can be embedded in its *full ring of fractions* $S^{-1}R$, where S consists of all non-divisors of 0 in R. An integral domain can be embedded in its fraction field.

Our goal is to study the relation between prime ideals of R and prime ideals of $S^{-1}R$.

1.2.2 Lemma

If X is any subset of R, define $S^{-1}X = \{x/s : x \in X, s \in S\}$. If I is an ideal of R, then $S^{-1}I$ is an ideal of $S^{-1}R$. If J is another ideal of R, then
(i) $S^{-1}(I + J) = S^{-1}I + S^{-1}J$;
(ii) $S^{-1}(IJ) = (S^{-1}I)(S^{-1}J)$;
(iii) $S^{-1}(I \cap J) = (S^{-1}I) \cap (S^{-1}J)$;
(iv) $S^{-1}I$ is a proper ideal iff $S \cap I = \emptyset$.

Proof. The definitions of addition and multiplication in $S^{-1}R$ imply that $S^{-1}R$ is an ideal, and that in (i), (ii) and (iii), the left side is contained in the right side. The reverse inclusions in (i) and (ii) follow from

$$\frac{a}{s} + \frac{b}{t} = \frac{at + bs}{st}, \quad \frac{a}{s}\frac{b}{t} = \frac{ab}{st}.$$

To prove (iii), let $a/s = b/t$, where $a \in I$, $b \in J$, $s, t \in S$. There exists $u \in S$ such that $u(at - bs) = 0$. Then $a/s = uat/ust = ubs/ust \in S^{-1}(I \cap J)$.

Finally, if $s \in S \cap I$, then $1/1 = s/s \in S^{-1}I$, so $S^{-1}I = S^{-1}R$. Conversely, if $S^{-1}I = S^{-1}R$, then $1/1 = a/s$ for some $a \in I$, $s \in S$. There exists $t \in S$ such that $t(s - a) = 0$, so $at = st \in S \cap I$. ♣

Ideals in $S^{-1}R$ must be of a special form.

1.2.3   Lemma

Let h be the natural homomorphism from R to $S^{-1}R$ [see (1.2.1)]. If J is an ideal of $S^{-1}R$ and $I = h^{-1}(J)$, then I is an ideal of R and $S^{-1}I = J$.

Proof. I is an ideal by the basic properties of preimages of sets. Let $a/s \in S^{-1}I$, with $a \in I$ and $s \in S$. Then $a/1 = h(a) \in J$, so $a/s = (a/1)(1/s) \in J$. Conversely, let $a/s \in J$, with $a \in R, s \in S$. Then $h(a) = a/1 = (a/s)(s/1) \in J$, so $a \in I$ and $a/s \in S^{-1}I$. ♣

Prime ideals yield sharper results.

1.2.4   Lemma

If I is any ideal of R, then $I \subseteq h^{-1}(S^{-1}I)$. There will be equality if I is prime and disjoint from S.

Proof. If $a \in I$, then $h(a) = a/1 \in S^{-1}I$. Thus assume that I is prime and disjoint from S, and let $a \in h^{-1}(S^{-1}I)$. Then $h(a) = a/1 \in S^{-1}I$, so $a/1 = b/s$ for some $b \in I, s \in S$. There exists $t \in S$ such that $t(as - b) = 0$. Thus $ast = bt \in I$, with $st \notin I$ because $S \cap I = \emptyset$. Since I is prime, we have $a \in I$. ♣

1.2.5   Lemma

If I is a prime ideal of R disjoint from S, then $S^{-1}I$ is a prime ideal of $S^{-1}R$.

Proof. By part (iv) of (1.2.2), $S^{-1}I$ is a proper ideal. Let $(a/s)(b/t) = ab/st \in S^{-1}I$, with $a, b \in R, s, t \in S$. Then $ab/st = c/u$ for some $c \in I, u \in S$. There exists $v \in S$ such that $v(abu - cst) = 0$. Thus $abuv = cstv \in I$, and $uv \notin I$ because $S \cap I = \emptyset$. Since I is prime, $ab \in I$, hence $a \in I$ or $b \in I$. Therefore either a/s or b/t belongs to $S^{-1}I$. ♣

The sequence of lemmas can be assembled to give a precise conclusion.

1.2.6   Theorem

There is a one-to-one correspondence between prime ideals P of R that are disjoint from S and prime ideals Q of $S^{-1}R$, given by

$$P \to S^{-1}P \text{ and } Q \to h^{-1}(Q).$$

Proof. By (1.2.3), $S^{-1}(h^{-1}(Q)) = Q$, and by (1.2.4), $h^{-1}(S^{-1}P) = P$. By (1.2.5), $S^{-1}P$ is a prime ideal, and $h^{-1}(Q)$ is a prime ideal by the basic properties of preimages of sets. If $h^{-1}(Q)$ meets S, then by (1.2.2) part (iv), $Q = S^{-1}(h^{-1}(Q)) = S^{-1}R$, a contradiction. Thus the maps $P \to S^{-1}P$ and $Q \to h^{-1}(Q)$ are inverses of each other, and the result follows. ♣

1.2.7 Definitions and Comments

If P is a prime ideal of R, then $S = R \setminus P$ is a multiplicative set. In this case, we write R_P for $S^{-1}R$, and call it the *localization* of R at P. We are going to show that R_P is a *local ring*, that is, a ring with a unique maximal ideal. First, we give some conditions equivalent to the definition of a local ring.

1.2.8 Proposition

For a ring R, the following conditions are equivalent.

(i) R is a local ring;
(ii) There is a proper ideal I of R that contains all nonunits of R;
(iii) The set of nonunits of R is an ideal.

Proof.
(i) implies (ii): If a is a nonunit, then (a) is a proper ideal, hence is contained in the unique maximal ideal I.
(ii) implies (iii): If a and b are nonunits, so are $a + b$ and ra. If not, then I contains a unit, so $I = R$, contradicting the hypothesis.
(iii) implies (i): If I is the ideal of nonunits, then I is maximal, because any larger ideal J would have to contain a unit, so $J = R$. If H is any proper ideal, then H cannot contain a unit, so $H \subseteq I$. Therefore I is the unique maximal ideal. ♣

1.2.9 Theorem

R_P is a local ring.

Proof. Let Q be a maximal ideal of R_P. Then Q is prime, so by (1.2.6), $Q = S^{-1}I$ for some prime ideal I of R that is disjoint from $S = R \setminus P$. In other words, $I \subseteq P$. Consequently, $Q = S^{-1}I \subseteq S^{-1}P$. If $S^{-1}P = R_P = S^{-1}R$, then by (1.2.2) part (iv), P is not disjoint from $S = R \setminus P$, which is impossible. Therefore $S^{-1}P$ is a proper ideal containing every maximal ideal, so it must be the unique maximal ideal. ♣

1.2.10 Remark

It is convenient to write the ideal $S^{-1}I$ as IR_P. There is no ambiguity, because the product of an element of I and an arbitrary element of R belongs to I.

1.2.11 Localization of Modules

If M is an R-module and S a multiplicative subset of R, we can essentially repeat the construction of (1.2.1) to form the localization of M by S, and thereby divide elements of M by elements of S. If $x, y \in M$ and $s, t \in S$, we call (x, s) and (y, t) equivalent if for some $u \in S$, we have $u(tx - sy) = 0$. The equivalence class of (x, s) is denoted by x/s, and addition is defined by

$$\frac{x}{s} + \frac{y}{t} = \frac{tx + sy}{st}.$$

If $a/s \in S^{-1}R$ and $x/t \in S^{-1}M$, we define

$$\frac{a}{s}\frac{x}{t} = \frac{ax}{st}.$$

In this way, $S^{-1}M$ becomes an $S^{-1}R$-module. Exactly as in (1.2.2), if M and N are submodules of an R-module L, then

$$S^{-1}(M + N) = S^{-1}M + S^{-1}N \text{ and } S^{-1}(M \cap N) = (S^{-1}M) \cap (S^{-1}N).$$

Problems For Section 1.2

1. Let $\mathcal{M}$ be a maximal ideal of R, and assume that for every $x \in \mathcal{M}$, $1 + x$ is a unit. Show that R is a local ring (with maximal ideal $\mathcal{M}$).

2. Show that if p is prime and n is a positive integer, then $\mathbb{Z}/p^n\mathbb{Z}$ is a local ring with maximal ideal (p).

3. For any field k, let R be the ring of rational functions f/g with $f, g \in k[X_1, \dots, X_n]$ and $g(a) \neq 0$, where a is a fixed point of k^n. Show that R is a local ring, and identify the unique maximal ideal.

Let S be a multiplicative subset of the ring R. We are going to construct a mapping from R-modules to $S^{-1}R$-modules, and another mapping from R-module homomorphisms to $S^{-1}R$-module homomorphisms, as follows. If M is an R-module, we map M to $S^{-1}M$. If $f : M \to N$ is an R-module homomorphism, we define $S^{-1}f : S^{-1}M \to S^{-1}N$ by

$$\frac{x}{s} \to \frac{f(x)}{s}.$$

Since f is a homomorphism, so is $S^{-1}f$. In Problems 4-6, we study these mappings.

4. Let $f : M \to N$ and $g : N \to L$ be R-module homomorphisms. Show that $S^{-1}(g \circ f) = (S^{-1}g) \circ (S^{-1}f)$. Also, if 1_M is the identity mapping on M, show that $S^{-1}1_M = 1_{S^{-1}M}$. Thus we have a functor S^{-1}, called the *localization functor*, from the category of R-modules to the category of $S^{-1}R$-modules.

5. If

$$M \xrightarrow{\quad f \quad} N \xrightarrow{\quad g \quad} L$$

is an exact sequence of R-modules, show that

$$S^{-1}M \xrightarrow{\quad S^{-1}f \quad} S^{-1}N \xrightarrow{\quad S^{-1}g \quad} S^{-1}L$$

is exact. Thus S^{-1} is an exact functor.

6. If M is an R-module and S is a multiplicative subset of R, denote $S^{-1}M$ by M_S. If N is a submodule of M, show that $(M/N)_S \cong M_S/N_S$.

Chapter 2

Norms, Traces and Discriminants

We continue building our algebraic background to prepare for algebraic number theory.

2.1 Norms and Traces

2.1.1 Definitions and Comments

If E/F is a field extension of finite degree n, then in particular, E is a finite-dimensional vector space over F, and the machinery of basic linear algebra becomes available. If x is any element of E, we can study the F-linear transformation $m(x)$ given by multiplication by x, that is, $m(x)y = xy$. We define the *norm* and the *trace* of x, relative to the extension E/F, as

$$N_{E/F}(x) = \det m(x) \text{ and } T_{E/F}(x) = \text{ trace } m(x).$$

We will write $N(x)$ and $T(x)$ if E/F is understood. If the matrix $A(x) = [a_{ij}(x)]$ represents $m(x)$ with respect to some basis for E over F, then the norm of x is the determinant of $A(x)$ and the trace of x is the trace of $A(x)$, that is, the sum of the main diagonal entries. The *characteristic polynomial* of x is defined as the characteristic polynomial of the matrix $A(x)$, that is,

$$\text{char}_{E/F}(x) = \det[XI - A(x)]$$

where I is an n by n identity matrix. It follows from the definitions that the norm, the trace and the coefficients of the characteristic polynomial are elements belonging to the base field F.

2.1.2 Example

Let $E = \mathbb{C}$ and $F = \mathbb{R}$. A basis for $\mathbb{C}$ over $\mathbb{R}$ is $\{1, i\}$ and, with $x = a + bi$, we have

$$(a + bi)(1) = a(1) + b(i) \text{ and } (a + bi)(i) = -b(1) + a(i).$$

9

Thus

$$A(a + bi) = \begin{bmatrix} a & -b \\ b & a \end{bmatrix}.$$

The norm, trace and characteristic polynomial of $a + bi$ are

$$N(a + bi) = a^2 + b^2, \; T(a + bi) = 2a, \; \text{char}(a + bi) = X^2 - 2aX + a^2 + b^2.$$

The computation is exactly the same if $E = \mathbb{Q}(i)$ and $F = \mathbb{Q}$.

2.1.3   Some Basic Properties

Notice that in (2.1.2), the coefficient of the second highest power of X in the characteristic polynomial is minus the trace, and the constant term is the norm. In general, it follows from the definition of characteristic polynomial that

$$\text{char}(x) = X^n - T(x)X^{n-1} + \cdots + (-1)^n N(x). \tag{1}$$

[The only terms multiplying X^{n-1} in the expansion of the determinant defining the characteristic polynomial are $-a_{ii}(x), i = 1, \dots, n$. Set $X = 0$ to show that the constant term of char(x) is $(-1)^n \det A(x)$.]

If $x, y \in E$ and $a, b \in F$, then

$$T(ax + by) = aT(x) + bT(y) \text{ and } N(xy) = N(x)N(y). \tag{2}$$

[This holds because $m(ax + by) = am(x) + bm(y)$ and $m(xy) = m(x) \circ m(y)$.]

If $a \in F$, then

$$N(a) = a^n, \; T(a) = na, \text{ and } \text{char}(a) = (X - a)^n. \tag{3}$$

[Note that the matrix representing multiplication by the element a in F is aI.]

It is natural to look for a connection between the characteristic polynomial of x and the minimal polynomial $\min(x, F)$ of x over F.

2.1.4   Proposition

$\text{char}_{E/F}(x) = [\min(x, F)]^r$, where $r = [E : F(x)]$.

Proof. First assume that $r = 1$, so that $E = F(x)$. By the Cayley-Hamilton theorem, the linear transformation $m(x)$ satisfies char(x). Since $m(x)$ is multiplication by x, it follows that x itself is a root of char(x). Thus $\min(x, F)$ divides char(x), and since both polynomials are monic of degree n, the result follows. In the general case, let $y_1, \dots, y_s$ be a basis for $F(x)$ over F, and let $z_1, \dots, z_r$ be a basis for E over $F(x)$. Then the $y_i z_j$ form a basis for E over F. Let $A = A(x)$ be the matrix representing multiplication by x in the extension $F(x)/F$, so that $xy_i = \sum_k a_{ki} y_k$ and $x(y_i z_j) = \sum_k a_{ki}(y_k z_j)$. Order the basis for E/F as $y_1 z_1, y_2 z_1, \ldots, y_s z_1; y_1 z_2, y_2 z_2 \ldots, y_s z_2; \cdots; y_1 z_r, y_2 z_r, \ldots, y_s z_r$. Then $m(x)$ is represented in E/F as

$$
\begin{bmatrix}
A & 0 & \cdots & 0 \\
0 & A & \cdots & 0 \\
\vdots & \vdots & & \vdots \\
0 & 0 & \cdots & A
\end{bmatrix}
$$

with r blocks, each consisting of the s by s matrix A. Thus $\mathrm{char}_{E/F}(x) = [\det(XI - A)]^r$, which by the $r = 1$ case coincides with $[\min(x, F)]^r$. ♣

2.1.5 Corollary

Let $[E : F] = n$ and $[F(x) : F] = d$. Let $x_1, \ldots, x_d$ be the roots of $\min(x, F)$, counting multiplicity, in a splitting field. Then

$$
N(x) = (\prod_{i=1}^{d} x_i)^{n/d}, \quad T(x) = \frac{n}{d} \sum_{i=1}^{d} x_i, \quad \mathrm{char}(x) = [\prod_{i=1}^{d}(X - x_i)]^{n/d}.
$$

Proof. The formula for the characteristic polynomial follows from (2.1.4). By (2.1.3), the norm is $(-1)^n$ times the constant term of $\mathrm{char}(x)$. Evaluating the characteristic polynomial at $X = 0$ produces another factor of $(-1)^n$, which yields the desired expression for the norm. Finally, if $\min(x, F) = X^d + a_{d-1} X^{d-1} + \cdots + a_1 X + a_0$, then the coefficient of X^{n-1} in $[\min(x, F)]^{n/d}$ is $(n/d) a_{d-1} = -(n/d) \sum_{i=1}^{d} x_i$. Since the trace is the negative of this coefficient [see (2.1.3)], the result follows. ♣

If E is a separable extension of F, there are very useful alternative expressions for the trace, norm and characteristic polynomial.

2.1.6 Proposition

Let E/F be a separable extension of degree n, and let $\sigma_1, \ldots, \sigma_n$ be the distinct F-embeddings (that is, F-monomorphisms) of E into an algebraic closure of E, or equally well into a normal extension L of F containing E. Then

$$
N_{E/F}(x) = \prod_{i=1}^{n} \sigma_i(x), \quad T_{E/F}(x) = \sum_{i=1}^{n} \sigma_i(x), \quad \mathrm{char}_{E/F}(x) = \prod_{i=1}^{n}(X - \sigma_i(x)).
$$

Proof. Each of the d distinct F-embeddings τ_i of $F(x)$ into L takes x into a unique conjugate x_i, and extends to exactly $n/d = [E : F(x)]$ F-embeddings of E into L, all of which also take x to x_i. Thus the list of elements $\{\sigma_1(x), \ldots, \sigma_n(x)\}$ consists of the $\tau_i(x) = x_i, i = 1, \ldots, d$, each appearing n/d times. The result follows from (2.1.5). ♣

We may now prove a basic transitivity property.

2.1.7   Transitivity of Trace and Norm

If $F \leq K \leq E$, where E/F is finite and separable, then

$$T_{E/F} = T_{K/F} \circ T_{E/K} \text{ and } N_{E/F} = N_{K/F} \circ N_{E/K}.$$

Proof. Let $\sigma_1, \ldots, \sigma_n$ be the distinct F-embeddings of K into L, and let $\tau_1, \ldots, \tau_m$ be the distinct K-embeddings of E into L, where L is the normal closure of E over F. Then L/F is Galois, and each mapping σ_i and τ_j extends to an automorphism of L. Therefore it makes sense to allow the mappings to be composed. By (2.1.6),

$$T_{K/F}(T_{E/K}(x)) = \sum_{i=1}^{n} \sigma_i \left(\sum_{j=1}^{m} \tau_j(x) \right) = \sum_{i=1}^{n} \sum_{j=1}^{m} \sigma_i(\tau_j(x)).$$

Each $\sigma_i \tau_j = \sigma_i \circ \tau_j$ is an F-embedding of E into L, and the number of mappings is given by $mn = [E : K][K : F] = [E : F]$. Furthermore, the $\sigma_i \tau_j$ are distinct when restricted to E. For if $\sigma_i \tau_j = \sigma_k \tau_l$ on E, then $\sigma_i = \sigma_k$ on K, because τ_j and τ_k coincide with the identity on K. Thus $i = k$, so that $\tau_j = \tau_l$ on E. But then $j = l$. By (2.1.6), $T_{K/F}(T_{E/K}(x)) = T_{E/F}(x)$. The norm is handled the same way, with sums replaced by products. ♣

Here is another application of (2.1.6).

2.1.8   Proposition

If E/F is a finite separable extension, then the trace $T_{E/F}(x)$ cannot be 0 for every $x \in E$.

Proof. If $T(x) = 0$ for all x, then by (2.1.6), $\sum_{i=1}^{n} \sigma_i(x) = 0$ for all x. This contradicts Dedekind's lemma on linear independence of monomorphisms. ♣

2.1.9   Remark

A statement equivalent to (2.1.8) is that if E/F is finite and separable, then the *trace form*, that is, the bilinear form $(x, y) \to T_{E/F}(xy)$, is nondegenerate. In other words, if $T(xy) = 0$ for all y, then $x = 0$. Going from (2.1.9) to (2.1.8) is immediate, so assume $T(xy) = 0$ for all y, with $x \neq 0$. Let x_0 be an element with nonzero trace, as provided by (2.1.8). Choose y so that $xy = x_0$ to produce a contradiction.

2.1.10   Example

Let $x = a + b\sqrt{m}$ be an element of the quadratic extension $\mathbb{Q}(\sqrt{m})/\mathbb{Q}$, where m is a square-free integer. We will find the trace and norm of x.

The Galois group of the extension consists of the identity and the automorphism $\sigma(a + b\sqrt{m}) = a - b\sqrt{m}$. Thus by (2.1.6),

$$T(x) = x + \sigma(x) = 2a, \text{ and } N(x) = x\sigma(x) = a^2 - mb^2.$$

Problems For Section 2.1

1. If $E = \mathbb{Q}(\theta)$ where θ is a root of the irreducible cubic $X^3 - 3X + 1$, find the norm and trace of θ^2.

2. Find the trace of the primitive 6th root of unity ω in the cyclotomic extension $\mathbb{Q}_6 = \mathbb{Q}(\omega)$.

3. Let θ be a root of $X^4 - 2$ over $\mathbb{Q}$. Find the trace over $\mathbb{Q}$ of $\theta, \theta^2, \theta^3$ and $\sqrt{3}\theta$.

4. Continuing Problem 3, show that $\sqrt{3}$ cannot belong to $\mathbb{Q}[\theta]$.

2.2 The Basic Setup For Algebraic Number Theory

2.2.1 Assumptions

Let A be an integral domain with fraction field K, and let L be a finite separable extension of K. Let B be the set of elements of L that are integral over A, that is, B is the integral closure of A in L. The diagram below summarizes the information.

In the most important special case, $A = \mathbb{Z}$, $K = \mathbb{Q}$, L is a *number field*, that is, a finite (necessarily separable) extension of $\mathbb{Q}$, and B is the ring of algebraic integers of L. From now on, we will refer to (2.2.1) as the *AKLB setup*.

2.2.2 Proposition

If $x \in B$, then the coefficients of $\text{char}_{L/K}(x)$ and $\min(x, K)$ are integral over A. In particular, $T_{L/K}(x)$ and $N_{L/K}(x)$ are integral over A, by (2.1.3). If A is integrally closed, then the coefficients belong to A.

Proof. The coefficients of $\min(x, K)$ are sums of products of the roots x_i, so by (2.1.4), it suffices to show that the x_i are integral over A. Each x_i is a conjugate of x over K, so there is a K-isomorphism $\tau_i : K(x) \to K(x_i)$ such that $\tau_i(x) = x_i$. If we apply τ_i to an equation of integral dependence for x over A, we get an equation of integral dependence for x_i over A. Since the coefficients belong to K [see (2.1.1)], they must belong to A if A is integrally closed. ♣

2.2.3 Corollary

Assume A integrally closed, and let $x \in L$. Then x is integral over A, that is, $x \in B$, if and only if the minimal polynomial of x over K has coefficients in A.

Proof. If $\min(x, K) \in A[X]$, then x is integral over A by definition of integrality. (See (1.1.1); note also that A need not be integrally closed for this implication.) The converse follows from (2.2.2). ♣

2.2.4   Corollary

An algebraic integer a that belongs to $\mathbb{Q}$ must in fact belong to $\mathbb{Z}$.

Proof. The minimal polynomial of a over $\mathbb{Q}$ is $X - a$, so by (2.2.3), $a \in \mathbb{Z}$. ♣

2.2.5   Quadratic Extensions of the Rationals

We will determine the algebraic integers of $L = \mathbb{Q}(\sqrt{m})$, where m is a square-free integer (a product of distinct primes). The restriction on m involves no loss of generality, for example, $\mathbb{Q}(\sqrt{12}) = \mathbb{Q}(\sqrt{3})$.

A remark on notation: To make sure there is no confusion between algebraic integers and ordinary integers, we will often use the term "rational integer" for a member of $\mathbb{Z}$.

Now by direct verification or by (2.1.10) and (2.1.3), the minimal polynomial over $\mathbb{Q}$ of the element $a + b\sqrt{m} \in L$ (with $a, b \in \mathbb{Q}$) is $X^2 - 2aX + a^2 - mb^2$. By (2.2.3), $a + b\sqrt{m}$ is an algebraic integer if and only if $2a$ and $a^2 - mb^2$ are rational integers. In this case, we also have $2b \in \mathbb{Z}$. For we have $(2a)^2 - m(2b)^2 = 4(a^2 - mb^2) \in \mathbb{Z}$, so $m(2b)^2 \in \mathbb{Z}$. If $2b$ is not a rational integer, its denominator would included a prime factor p, which would appear as p^2 in the denominator of $(2b)^2$. Multiplication of $(2b)^2$ by m cannot cancel the p^2 because m is square-free, and the result follows.

Here is a more convenient way to characterize the algebraic integers of a quadratic field.

2.2.6   Proposition

The set B of algebraic integers of $\mathbb{Q}(\sqrt{m})$, m square-free, can be described as follows.

(i) If $m \not\equiv 1 \mod 4$, then B consists of all $a + b\sqrt{m}$, $a, b \in \mathbb{Z}$;

(ii) If $m \equiv 1 \mod 4$, then B consists of all $(u/2) + (v/2)\sqrt{m}$, $u, v \in \mathbb{Z}$, where u and v have the same parity (both even or both odd).

[Note that since m is square-free, it is not divisible by 4, so the condition in (i) can be written as $m \equiv 2$ or $3 \mod 4$.]

Proof. By (2.2.5), the algebraic integers are of the form $(u/2) + (v/2)\sqrt{m}$, where $u, v \in \mathbb{Z}$ and $(u^2 - mv^2)/4 \in \mathbb{Z}$, that is, $u^2 - mv^2 \equiv 0 \mod 4$. It follows that u and v have the same parity. [The square of an even number is congruent to 0 mod 4, and the square of an odd number is congruent to 1 mod 4.] Moreover, the "both odd" case can only occur when $m \equiv 1 \mod 4$. The "both even" case is equivalent to $u/2, v/2 \in \mathbb{Z}$, and we have the desired result. ♣

When we introduce integral bases in the next section, we will have an even more convenient way to describe the algebraic integers of $\mathbb{Q}(\sqrt{m})$.

If $[L : K] = n$, then a basis for L/K consists of n elements of L that are linearly independent over K. In fact we can assemble a basis using only elements of B.

2.2.7   Proposition

There is a basis for L/K consisting entirely of elements of B.

Proof. Let $x_1, \ldots, x_n$ be a basis for L over K. Each x_i is algebraic over K, and therefore satisfies a polynomial equation of the form

$$a_m x_i^m + \cdots + a_1 x_i + a_0 = 0$$

with $a_m \neq 0$ and the $a_i \in A$. (Initially, we only have $a_i \in K$, but then a_i is the ratio of two elements of A, and we can form a common denominator.) Multiply the equation by a_m^{m-1} to obtain an equation of integral dependence for $y_i = a_m x_i$ over A. The y_i form the desired basis. ♣

2.2.8 Corollary of the Proof

If $x \in L$, then there is a nonzero element $a \in A$ and an element $y \in B$ such that $x = y/a$. In particular, L is the fraction field of B.

Proof. In the proof of (2.2.7), take $x_i = x$, $a_m = a$, and $y_i = y$. ♣

In Section 2.3, we will need a standard result from linear algebra. We state the result now, and an outline of the proof is given in the exercises.

2.2.9 Theorem

Suppose we have a nondegenerate symmetric bilinear form on an n-dimensional vector space V, written for convenience using inner product notation (x, y). If $x_1, \ldots, x_n$ is any basis for V, then there is a basis $y_1, \ldots, y_n$ for V, called the *dual basis referred to V*, such that

$$(x_i, y_j) = \delta_{ij} = \begin{cases} 1, & i = j \\ 0, & i \neq j. \end{cases}$$

Problems For Section 2.2

1. Let $L = \mathbb{Q}(\alpha)$, where α is a root of the irreducible quadratic $X^2 + bX + c$, with $b, c \in \mathbb{Q}$. Show that $L = \mathbb{Q}(\sqrt{m})$ for some square-free integer m. Thus the analysis of this section covers all possible quadratic extensions of $\mathbb{Q}$.
2. Show that the quadratic extensions $\mathbb{Q}(\sqrt{m})$, m square-free, are all distinct.
3. Continuing Problem 2, show that in fact no two distinct quadratic extensions of $\mathbb{Q}$ are $\mathbb{Q}$-isomorphic.

Cyclotomic fields do not exhibit the same behavior. Let $\omega_n = e^{i2\pi/n}$, a primitive n^{th} root of unity. By a direct computation, we have $\omega_{2n}^2 = \omega_n$ and

$$-\omega_{2n}^{n+1} = -e^{i\pi(n+1)/n} = e^{i\pi} e^{i\pi} e^{i\pi/n} = \omega_{2n}.$$

4. Show that if n is odd, then $\mathbb{Q}(\omega_n) = \mathbb{Q}(\omega_{2n})$.
5. Give an example of a quadratic extension of $\mathbb{Q}$ that is also a cyclotomic extension.

We now indicate how to prove (2.2.9).
6. For any y in the finite-dimensional vector space V, the mapping $x \to (x, y)$ is a linear form $l(y)$ on V, that is, a linear map from V to the field of scalars. Show that the linear

transformation $y \to l(y)$ from V to V^* (the space of all linear forms on V) is injective.

7. Show that any linear form on V is $l(y)$ for some y.

8. Let $f_1, \ldots, f_n$ be the dual basis corresponding to $x_1, \ldots, x_n$. Thus each f_j belongs to V^* (*not* V) and $f_j(x_i) = \delta_{ij}$. If $f_j = l(y_j)$, show that $y_1, \ldots, y_n$ is the required dual basis referred to V.

9. Show that $x_i = \sum_{j=1}^n (x_i, x_j) y_j$. Thus in order to compute the dual basis referred to V, we must invert the matrix $((x_i, x_j))$.

2.3   The Discriminant

The discriminant of a polynomial is familiar from basic algebra, and there is also a discriminant in algebraic number theory. The two concepts are unrelated at first glance, but there is a connection between them. We assume the basic *AKLB* setup of (2.2.1), with $n = [L : K]$.

2.3.1   Definition

If $n = [L : K]$, the *discriminant* of the n-tuple $x = (x_1, \ldots, x_n)$ of elements of L is

$$D(x) = \det(T_{L/K}(x_i x_j)).$$

Thus we form a matrix whose ij entry is the trace of $x_i x_j$, and take the determinant of the matrix; by (2.1.1), $D(x) \in K$. If the $x_i \in B$, then by (2.2.2), $D(x)$ is integral over A, that is, $D(x) \in B$. Thus if A is integrally closed and the $x_i \in B$, then $D(x)$ belongs to A.

The discriminant behaves quite reasonably under linear transformation.

2.3.2   Lemma

If $y = Cx$, where C is an n by n matrix over K and x and y are n-tuples written as column vectors, then $D(y) = (\det C)^2 D(x)$.

Proof. The trace of $y_r y_s$ is

$$T\left(\sum_{i,j} c_{ri} c_{sj} x_i x_j\right) = \sum_{i,j} c_{ri} T(x_i x_j) c_{sj}$$

hence

$$(T(y_r y_s)) = C(T(x_i x_j)) C'$$

where C' is the transpose of C. The result follows upon taking determinants. ♣

Here is an alternative expression for the discriminant.

2.3.3   Lemma

Let $\sigma_1, \ldots, \sigma_n$ be the distinct K-embeddings of L into an algebraic closure of L, as in (2.1.6). Then $D(x) = [\det(\sigma_i(x_j))]^2$.

Thus we form the matrix whose ij element is $\sigma_i(x_j)$, take the determinant and square the result.

Proof. By (2.1.6),

$$T(x_i x_j) = \sum_k \sigma_k(x_i x_j) = \sum_k \sigma_k(x_i)\sigma_k(x_j)$$

so if H is the matrix whose ij entry is $\sigma_i(x_j)$, then $(T(x_i x_j)) = H'H$, and again the result follows upon taking determinants. ♣

The discriminant "discriminates" between bases and non-bases, as follows.

2.3.4 Proposition

If $x = (x_1, \ldots, x_n)$, then the x_i form a basis for L over K if and only if $D(x) \neq 0$.

Proof. If $\sum_j c_j x_j = 0$, with the $c_j \in K$ and not all 0, then $\sum_j c_j \sigma_i(x_j) = 0$ for all i, so the columns of the matrix $H = (\sigma_i(x_j))$ are linearly dependent. Thus linear dependence of the x_i implies that $D(x) = 0$. Conversely, assume that the x_i are linearly independent, and therefore a basis because $n = [L : K]$. If $D(x) = 0$, then the rows of H are linearly dependent, so for some $c_i \in K$, not all 0, we have $\sum_i c_i \sigma_i(x_j) = 0$ for all j. Since the x_j form a basis, it follows that $\sum_i c_i \sigma_i(u) = 0$ for all $u \in L$, so the monomorphisms σ_i are linearly dependent. This contradicts Dedekind's lemma. ♣

We now make the connection between the discriminant defined above and the discriminant of a polynomial.

2.3.5 Proposition

Assume that $L = K(x)$, and let f be the minimal polynomial of x over K. Let D be the discriminant of the basis $1, x, x^2, \ldots, x^{n-1}$ over K, and let $x_1, \ldots, x_n$ be the roots of f in a splitting field, with $x_1 = x$. Then D coincides with $\prod_{i<j}(x_i - x_j)^2$, the discriminant of the polynomial f.

Proof. Let σ_i be the K-embedding that takes x to x_i, $i = 1, \ldots, n$. Then $\sigma_i(x^j) = x_i^j$, $0 \leq j \leq n - 1$. By (2.3.3), D is the square of the determinant of the matrix

$$V = \begin{bmatrix} 1 & x_1 & x_1^2 & \cdots & x_1^{n-1} \\ 1 & x_2 & x_2^2 & \cdots & x_2^{n-1} \\ \vdots & \vdots & \vdots & \ddots & \vdots \\ 1 & x_n & x_n^2 & \cdots & x_n^{n-1} \end{bmatrix}.$$

But $\det V$ is a Vandermonde determinant, whose value is $\prod_{i<j}(x_j - x_i)$, and the result follows. ♣

Proposition 2.3.5 yields a formula that is often useful in computing the discriminant.

2.3.6   Corollary

Under the hypothesis of (2.3.5),

$$D = (-1)^{\binom{n}{2}} N_{L/K}(f'(x))$$

where f' is the derivative of f.

Proof. Let $c = (-1)^{\binom{n}{2}}$. By (2.3.5),

$$D = \prod_{i<j}(x_i - x_j)^2 = c\prod_{i \neq j}(x_i - x_j) = c\prod_i \prod_{j \neq i}(x_i - x_j).$$

But $f(X) = (X - x_1)\cdots(X - x_n)$, so

$$f'(x_i) = \sum_k \prod_{j \neq k}(X - x_j)$$

with X replaced by x_i. When the substitution $X = x_i$ is carried out, only the $k = i$ term is nonzero, hence

$$f'(x_i) = \prod_{j \neq i}(x_i - x_j).$$

Consequently,

$$D = c\prod_{i=1}^{n} f'(x_i).$$

But

$$f'(x_i) = f'(\sigma_i(x)) = \sigma_i(f'(x))$$

so by (2.1.6),

$$D = cN_{L/K}(f'(x)). \quad \clubsuit$$

2.3.7   Definitions and Comments

In the $AKLB$ setup with $[L : K] = n$, suppose that B turns out to be a free A-module of rank n. A basis for this module is said to be an *integral basis* of B (or of L). An integral basis is, in particular, a basis for L over K, because linear independence over A is equivalent to linear independence over the fraction field K. We will see shortly that an integral basis always exists when L is a number field. In this case, the discriminant is the same for all integral bases. It is called the *field discriminant*.

2.3.8   Theorem

If A is integrally closed, then B is a submodule of a free A-module of rank n. If A is a PID, then B itself is free of rank n over A, so B has an integral basis.

Proof. By (2.1.9), the trace is a nondegenerate symmetric bilinear form defined on the n-dimensional vector space L over K. By (2.2.2), the trace of any element of B belongs to A. Now let $x_1, \ldots, x_n$ be any basis for L over K consisting of elements of B [see (2.2.7)], and let $y_1, \ldots, y_n$ be the dual basis referred to L [see (2.2.9)]. If $z \in B$, then we can write $z = \sum_{j=1} a_j y_j$ with the $a_j \in K$. We know that the trace of $x_i z$ belongs to A, and we also have

$$T(x_i z) = T(\sum_{j=1}^{n} a_j x_i y_j) = \sum_{j=1}^{n} a_j T(x_i y_j) = \sum_{j=1}^{n} a_j \delta_{ij} = a_i.$$

Thus each a_i belongs to A, so that B is an A-submodule of the free A-module $\oplus_{j=1}^{n} A y_j$. Moreover, B contains the free A-module $\oplus_{j=1}^{n} A x_j$. Consequently, if A is a principal ideal domain, then B is free over A of rank exactly n. ♣

2.3.9   Corollary

The set B of algebraic integers in any number field L is a free $\mathbb{Z}$-module of rank $n = [L : \mathbb{Q}]$. Therefore B has an integral basis. The discriminant is the same for every integral basis.

Proof. Take $A = \mathbb{Z}$ in (2.3.8) to show that B has an integral basis. The transformation matrix C between two integral bases [see (2.3.2)] is invertible, and both C and C^{-1} have rational integer coefficients. Take determinants in the equation $CC^{-1} = I$ to conclude that $\det C$ is a unit in $\mathbb{Z}$. Therefore $\det C = \pm 1$, so by (2.3.2), all integral bases have the same discriminant. ♣

2.3.10   Remark

An invertible matrix C with coefficients in $\mathbb{Z}$ is said to be *unimodular* if C^{-1} also has coefficients in $\mathbb{Z}$. We have just seen that a unimodular matrix has determinant ± 1. Conversely, a matrix over $\mathbb{Z}$ with determinant ± 1 is unimodular, by Cramer's rule.

2.3.11   Theorem

Let B be the algebraic integers of $\mathbb{Q}(\sqrt{m})$, where m is a square-free integer.

(i) If $m \not\equiv 1 \mod 4$, then 1 and $\sqrt{m}$ form an integral basis, and the field discriminant is $d = 4m$.

(ii) If $m \equiv 1 \mod 4$, then 1 and $(1 + \sqrt{m})/2$ form an integral basis, and the field discriminant is $d = m$.

Proof.

(i) By (2.2.6), 1 and $\sqrt{m}$ span B over $\mathbb{Z}$, and they are linearly independent because $\sqrt{m}$ is irrational. By (2.1.10), the trace of $a + b\sqrt{m}$ is $2a$, so by (2.3.1), the field discriminant is

$$\begin{vmatrix} 2 & 0 \\ 0 & 2m \end{vmatrix} = 4m.$$

(ii) By (2.2.6), 1 and $(1 + \sqrt{m})/2$ are algebraic integers. To show that they span B, consider $(u + v\sqrt{m})/2$, where u and v have the same parity. Then

$$\frac{1}{2}(u + v\sqrt{m}) = (\frac{u - v}{2})(1) + v \, [\frac{1}{2}(1 + \sqrt{m})]$$

with $(u - v)/2$ and v in $\mathbb{Z}$. To prove linear independence, assume that $a, b \in \mathbb{Z}$ and

$$a + b \, [\frac{1}{2}(1 + \sqrt{m})] = 0.$$

Then $2a + b + b\sqrt{m} = 0$, which forces $a = b = 0$. Finally, by (2.1.10), (2.3.1), and the computation $[(1 + \sqrt{m})/2]^2 = (1 + m)/4 + \sqrt{m}/2$, the field discriminant is

$$\begin{vmatrix} 2 & 1 \\ 1 & (1 + m)/2 \end{vmatrix} = m. \; \clubsuit$$

Problems For Section 2.3

Problems 1-3 outline the proof of *Stickelberger's theorem*, which states that the discriminant of any n-tuple in a number field is congruent to 0 or 1 mod 4.

1. Let $x_1, \ldots, x_n$ be arbitrary algebraic integers in a number field, and consider the determinant of the matrix $(\sigma_i(x_j))$, as in (2.3.3). The direct expansion of the determinant has $n!$ terms. let P be the sum of those terms in the expansion that have plus signs in front of them, and N the sum of those terms prefixed by minus signs. Thus the discriminant D of $(x_1, \ldots, x_n)$ is $(P - N)^2$. Show that $P + N$ and PN are fixed by each σ_i, and deduce that $P + N$ and PN are rational numbers.

2. Show that $P + N$ and PN are rational integers.

3. Show that $D \equiv 0$ or 1 mod 4.

4. Let L be a number field of degree n over $\mathbb{Q}$, and let $y_1, \ldots, y_n$ be a basis for L over $\mathbb{Q}$ consisting of algebraic integers. Let $x_1, \ldots, x_n$ be an integral basis. Show that if the discriminant $D(y_1, \ldots, y_n)$ is square-free, then each x_i can be expressed as a linear combination of the y_j with integer coefficients.

5. Continuing Problem 4, show that if $D(y_1, \ldots, y_n)$ is square-free, then $y_1, \ldots, y_n$ is an integral basis.

6. Is the converse of the result of problem 5 true?

Chapter 3

Dedekind Domains

3.1 The Definition and Some Basic Properties

We identify the natural class of integral domains in which unique factorization of ideals is possible.

3.1.1 Definition

A *Dedekind domain* is an integral domain A satisfying the following three conditions:
(1) A is a Noetherian ring;
(2) A is integrally closed;
(3) Every nonzero prime ideal of A is maximal.

A principal ideal domain satisfies all three conditions, and is therefore a Dedekind domain. We are going to show that in the $AKLB$ setup, if A is a Dedekind domain, then so is B, a result that provides many more examples and already suggests that Dedekind domains are important in algebraic number theory.

3.1.2 Proposition

In the $AKLB$ setup, B is integrally closed, regardless of A. If A is an integrally closed Noetherian ring, then B is also a Noetherian ring, as well as a finitely generated A-module.

Proof. By (1.1.6), B is integrally closed in L, which is the fraction field of B by (2.2.8). Therefore B is integrally closed. If A is integrally closed, then by (2.3.8), B is a submodule of a free A-module M of rank n. If A is Noetherian, then M, which is isomorphic to the direct sum of n copies of A, is a Noetherian A-module, hence so is the submodule B. An ideal of B is, in particular, an A-submodule of B, hence is finitely generated over A and therefore over B. It follows that B is a Noetherian ring. ♣

3.1.3 Theorem

In the $AKLB$ setup, if A is a Dedekind domain, then so is B. In particular, the ring of algebraic integers in a number field is a Dedekind domain.

21

Proof. In view of (3.1.2), it suffices to show that every nonzero prime ideal Q of B is maximal. Choose any nonzero element x of Q. Since $x \in B$, x satisfies a polynomial equation

$$x^m + a_{m-1}x^{m-1} + \cdots + a_1 x + a_0 = 0$$

with the $a_i \in A$. If we take the positive integer m as small as possible, then $a_0 \neq 0$ by minimality of m. Solving for a_0, we see that $a_0 \in Bx \cap A \subseteq Q \cap A$, so the prime ideal $P = Q \cap A$ is nonzero, hence maximal by hypothesis. By Section 1.1, Problem 6, Q is maximal. ♣

Problems For Section 3.1

This problem set will give the proof of a result to be used later. Let $P_1, P_2, \ldots, P_s$, $s \geq 2$, be ideals in a ring R, with P_1 and P_2 not necessarily prime, but $P_3, \ldots, P_s$ prime (if $s \geq 3$). Let I be any ideal of R. The idea is that if we can avoid the P_j individually, in other words, for each j we can find an element in I but not in P_j, then we can avoid all the P_j simultaneously, that is, we can find a single element in I that is in none of the P_j. The usual statement is the contrapositive of this assertion.

Prime Avoidance Lemma

With I and the P_i as above, if $I \subseteq \cup_{i=1}^{s} P_i$, then for some i we have $I \subseteq P_i$.

1. Suppose that the result is false. Show that without loss of generality, we can assume the existence of elements $a_i \in I$ with $a_i \in P_i$ but $a_i \notin P_1 \cup \cdots \cup P_{i-1} \cup P_{i+1} \cup \cdots \cup P_s$.
2. Prove the result for $s = 2$.
3. Now assume $s > 2$, and observe that $a_1 a_2 \cdots a_{s-1} \in P_1 \cap \cdots \cap P_{s-1}$, but $a_s \notin P_1 \cup \cdots \cup P_{s-1}$. Let $a = (a_1 \cdots a_{s-1}) + a_s$, which does not belong to $P_1 \cup \cdots \cup P_{s-1}$, else a_s would belong to this set. Show that $a \in I$ and $a \notin P_1 \cup \cdots \cup P_s$, contradicting the hypothesis.

3.2 Fractional Ideals

Our goal is to establish unique factorization of ideals in a Dedekind domain, and to do this we will need to generalize the notion of ideal. First, some preliminaries.

3.2.1 Products of Ideals

Recall that if $I_1, \ldots, I_n$ are ideals, then their product $I_1 \cdots I_n$ is the set of all finite sums $\sum_i a_{1i} a_{2i} \cdots a_{ni}$, where $a_{ki} \in I_k$, $k = 1, \ldots, n$. It follows from the definition that $I_1 \cdots I_n$ is an ideal contained in each I_j. Moreover, if a prime ideal P contains a product $I_1 \cdots I_n$ of ideals, then P contains I_j for some j.

3.2.2 Proposition

If I is a nonzero ideal of the Noetherian integral domain R, then I contains a product of nonzero prime ideals.

Proof. Assume the contrary. If $\mathcal{S}$ is the collection of all nonzero ideals that do not contain a product of nonzero prime ideals, then, as R is Noetherian, $\mathcal{S}$ has a maximal element J, and J cannot be prime because it belongs to $\mathcal{S}$. Thus there are elements $a, b \in R$ such that $a \notin J$, $b \notin J$, and $ab \in J$. By maximality of J, the ideals $J + Ra$ and $J + Rb$ each contain a product of nonzero prime ideals, hence so does $(J + Ra)(J + Rb) \subseteq J + Rab = J$. This is a contradiction. (Notice that we must use the fact that a product of nonzero ideals is nonzero, and this is where the hypothesis that R is an integral domain comes in.) ♣

3.2.3 Corollary

If I is an ideal of the Noetherian ring R (not necessarily an integral domain), then I contains a product of prime ideals.

Proof. Repeat the proof of (3.2.2), with the word "nonzero" deleted. ♣

Ideals in the ring of integers are of the form $n\mathbb{Z}$, the set of multiples of n. A set of the form $(3/2)\mathbb{Z}$ is not an ideal because it is not a subset of $\mathbb{Z}$, yet it behaves in a similar manner. The set is closed under addition and multiplication by an integer, and it becomes an ideal of $\mathbb{Z}$ if we simply multiply all the elements by 2. It will be profitable to study sets of this type.

3.2.4 Definitions

Let R be an integral domain with fraction field K, and let I be an R-submodule of K. We say that I is a *fractional ideal* of R if $rI \subseteq R$ for some nonzero $r \in R$. We call r a *denominator* of I. An ordinary ideal of R is a fractional ideal (take $r = 1$), and will often be referred to as an *integral ideal*.

3.2.5 Lemma

(i) If I is a finitely generated R-submodule of K, then I is a fractional ideal.

(ii) If R is Noetherian and I is a fractional ideal of R, then I is a finitely generated R-submodule of K.

(iii) If I and J are fractional ideals with denominators r and s respectively, then $I \cap J$, $I + J$ and IJ are fractional ideals with respective denominators r (or s), rs and rs. [The product of fractional ideals is defined exactly as in (3.2.1).]

Proof.

(i) If $x_1 = a_1/b_1, \dots, x_n = a_n/b_n$ generate I and $b = b_1 \cdots b_n$, then $bI \subseteq R$.

(ii) If $rI \subseteq R$, then $I \subseteq r^{-1}R$. As an R-module, $r^{-1}R$ is isomorphic to R and is therefore Noetherian. Consequently, I is finitely generated.

(iii) It follows from the definition (3.2.4) that the intersection, sum and product of fractional ideals are fractional ideals. The assertions about denominators are proved by noting that $r(I \cap J) \subseteq rI \subseteq R$, $rs(I + J) \subseteq rI + sJ \subseteq R$, and $rsIJ = (rI)(sJ) \subseteq R$. ♣

The product of two nonzero fractional ideals is a nonzero fractional ideal, and the multiplication is associative because multiplication in R is associative. There is an identity element, namely R, since $RI \subseteq I = 1I \subseteq RI$. We will show that if R is a Dedekind domain, then every nonzero fractional ideal has a multiplicative inverse, so the nonzero fractional ideals form a group.

3.2.6   Lemma

Let I be a nonzero prime ideal of the Dedekind domain R, and let J be the set of all elements $x \in K$ such that $xI \subseteq R$. Then $R \subset J$.

Proof. Since $RI \subseteq R$, it follows that R is a subset of J. Pick a nonzero element $a \in I$, so that I contains the principal ideal Ra. Let n be the smallest positive integer such that Ra contains a product $P_1 \cdots P_n$ of n nonzero prime ideals. Since R is Noetherian, there is such an n by (3.2.2), and by (3.2.1), I contains one of the P_i, say P_1. But in a Dedekind domain, every nonzero prime ideal is maximal, so $I = P_1$. Assuming $n \geq 2$, set $I_1 = P_2 \cdots P_n$, so that $Ra \not\supseteq I_1$ by minimality of n. Choose $b \in I_1$ with $b \notin Ra$. Now $II_1 = P_1 \cdots P_n \subseteq Ra$, in particular, $Ib \subseteq Ra$, hence $Iba^{-1} \subseteq R$. (Note that a has an inverse in K but not necessarily in R.) Thus $ba^{-1} \in J$, but $ba^{-1} \notin R$, for if so, $b \in Ra$, contradicting the choice of b.

The case $n = 1$ must be handled separately. In this case, $P_1 = I \supseteq Ra \supseteq P_1$, so $I = Ra$. Thus Ra is a proper ideal, and we can choose $b \in R$ with $b \notin Ra$. Then $ba^{-1} \notin R$, but $ba^{-1}I = ba^{-1}Ra = bR \subseteq R$, so $ba^{-1} \in J$. ♣

We now prove that in (3.2.6), J is the inverse of I.

3.2.7   Proposition

Let I be a nonzero prime ideal of the Dedekind domain R, and let $J = \{x \in K : xI \subseteq R\}$. Then J is a fractional ideal and $IJ = R$.

Proof. If r is a nonzero element of I and $x \in J$, then $rx \in R$, so $rJ \subseteq R$ and J is a fractional ideal. Now $IJ \subseteq R$ by definition of J, so IJ is an integral ideal. Using (3.2.6), we have $I = IR \subseteq IJ \subseteq R$, and maximality of I implies that either $IJ = I$ or $IJ = R$. In the latter case, we are finished, so assume $IJ = I$.

If $x \in J$, then $xI \subseteq IJ = I$, and by induction, $x^nI \subseteq I$ for all $n = 1, 2, \ldots$. Let r be any nonzero element of I. Then $rx^n \in x^nI \subseteq I \subseteq R$, so $R[x]$ is a fractional ideal. Since R is Noetherian, part (ii) of (3.2.5) implies that $R[x]$ is a finitely generated R-submodule of K. By (1.1.2), x is integral over R. But R, a Dedekind domain, is integrally closed, so $x \in R$. Therefore $J \subseteq R$, contradicting (3.2.6). ♣

The following basic property of Dedekind domains can be proved directly from the definition, without waiting for the unique factorization of ideals.

3.2.8   Theorem

If R is a Dedekind domain, then R is a UFD if and only if R is a PID.

Proof. Recall from basic algebra that a (commutative) ring R is a PID iff R is a UFD and every nonzero prime ideal of R is maximal. ♣

Problems For Section 3.2

1. If I and J are relatively prime ideals $(I + J = R)$, show that $IJ = I \cap J$. More generally, if $I_1, \ldots, I_n$ are relatively prime in pairs, show that $I_1 \cdots I_n = \cap_{i=1}^n I_i$.

2. Let P_1 and P_2 be relatively prime ideals in the ring R. Show that P_1^r and P_2^s are relatively prime for arbitrary positive integers r and s.

3. Let R be an integral domain with fraction field K. If K is a fractional ideal of R, show that $R = K$.

3.3   Unique Factorization of Ideals

In the previous section, we inverted nonzero prime ideals in a Dedekind domain. We now extend this result to nonzero fractional ideals.

3.3.1   Theorem

If I is a nonzero fractional ideal of the Dedekind domain R, then I can be factored uniquely as $P_1^{n_1} P_2^{n_2} \cdots P_r^{n_r}$, where the n_i are integers. Consequently, the nonzero fractional ideals form a group under multiplication.

Proof. First consider the existence of such a factorization. Without loss of generality, we can restrict to integral ideals. [Note that if $r \neq 0$ and $rI \subseteq R$, then $I = (rR)^{-1}(rI)$.] By convention, we regard R as the product of the empty collection of prime ideals, so let $\mathcal{S}$ be the set of all nonzero proper ideals of R that cannot be factored in the given form, with all n_i *positive* integers. (This trick will yield the useful result that the factorization of integral ideals only involves positive exponents.) Since R is Noetherian, $\mathcal{S}$, if nonempty, has a maximal element I_0, which is contained in a maximal ideal I. By (3.2.7), I has an inverse fractional ideal J. Thus by (3.2.6) and (3.2.7),

$$I_0 = I_0 R \subseteq I_0 J \subseteq IJ = R.$$

Therefore $I_0 J$ is an integral ideal, and we claim that $I_0 \subset I_0 J$. For if $I_0 = I_0 J$, then the last paragraph of the proof of (3.2.7) can be reproduced with I replaced by I_0 to reach a contradiction. By maximality of I_0, $I_0 J$ is a product of prime ideals, say $I_0 J = P_1 \cdots P_r$ (with repetition allowed). Multiply both sides by the prime ideal I to conclude that I_0 is a product of prime ideals, contradicting $I_0 \in \mathcal{S}$. Thus $\mathcal{S}$ must be empty, and the existence of the desired factorization is established.

To prove uniqueness, suppose that we have two prime factorizations

$$P_1^{n_1} \cdots P_r^{n_r} = Q_1^{t_1} \cdots Q_s^{t_s}$$

where again we may assume without loss of generality that all exponents are positive. (If P^{-n} appears, multiply both sides by P^n.) Now P_1 contains the product of the $P_i^{n_i}$, so by (3.2.1), P_1 contains Q_j for some j. By maximality of Q_j, $P_1 = Q_j$, and we may renumber so that $P_1 = Q_1$. Multiply by the inverse of P_1 (a fractional ideal, but there is no problem), and continue inductively to complete the proof. ♣

3.3.2   Corollary

A nonzero fractional ideal I is an integral ideal if and only if all exponents in the prime factorization of I are nonnegative.

Proof. The "only if" part was noted in the proof of (3.3.1). The "if" part follows because a power of an integral ideal is still an integral ideal. ♣

3.3.3   Corollary

Denote by $n_P(I)$ the exponent of the prime ideal P in the factorization of I. (If P does not appear, take $n_P(I) = 0$.) If I_1 and I_2 are nonzero fractional ideals, then $I_1 \supseteq I_2$ if and only if for every prime ideal P of R, $n_P(I_1) \leq n_P(I_2)$.

Proof. We have $I_2 \subseteq I_1$ iff $I_2 I_1^{-1} \subseteq R$, and by (3.3.2), this happens iff for every P, $n_P(I_2) - n_P(I_1) \geq 0$. ♣

3.3.4   Definition

Let I_1 and I_2 be nonzero integral ideals. We say that I_1 *divides* I_2 if $I_2 = JI_1$ for some integral ideal J. Just as with integers, an equivalent statement is that each prime factor of I_1 is a factor of I_2.

3.3.5   Corollary

If I_1 and I_2 are nonzero integral ideals, then I_1 divides I_2 if and only if $I_1 \supseteq I_2$. In other words, for these ideals,

$$\boxed{DIVIDES\ \ MEANS\ \ CONTAINS.}$$

Proof. By (3.3.4), I_1 divides I_2 iff $n_P(I_1) \leq n_P(I_2)$ for every prime ideal P. By (3.3.3), this is equivalent to $I_1 \supseteq I_2$. ♣

3.3.6   GCD's and LCM's

As a nice application of the principle that divides means contains, we can use the prime factorization of ideals in a Dedekind domain to compute the greatest common divisor and least common multiple of two nonzero ideals I and J, exactly as with integers. The greatest common divisor is the smallest ideal containing both I and J, that is, $I + J$. The least common multiple is the largest ideal contained in both I and J, which is $I \cap J$.

A Dedekind domain comes close to being a principal ideal domain in the sense that every nonzero integral ideal, in fact every nonzero fractional ideal, divides some principal ideal.

3.3.7 Proposition

Let I be a nonzero fractional ideal of the Dedekind domain R. Then there is a nonzero integral ideal J such that IJ is a principal ideal of R.

Proof. By (3.3.1), there is a nonzero fractional ideal I' such that $II' = R$. By definition of fractional ideal, there is a nonzero element $r \in R$ such that rI' is an integral ideal. If $J = rI'$, then $IJ = Rr$, a principal ideal of R. ♣

Problems For Section 3.3

By (2.3.11), the ring B of algebraic integers in $\mathbb{Q}(\sqrt{-5})$ is $\mathbb{Z}[\sqrt{-5}]$. In Problems 1-3, we will show that $\mathbb{Z}[\sqrt{-5}]$ is not a unique factorization domain by considering the factorization

$$(1 + \sqrt{-5})(1 - \sqrt{-5}) = 2 \times 3.$$

1. By computing norms, verify that all four of the above factors are irreducible.
2. Show that the only units of B are ± 1.
3. Show that no factor on one side of the above equation is an associate of a factor on the other side, so unique factorization fails.
4. Show that the ring of algebraic integers in $\mathbb{Q}(\sqrt{-17})$ is not a unique factorization domain.
5. In $\mathbb{Z}[\sqrt{-5}]$ and $\mathbb{Z}\sqrt{-17}]$, the only algebraic integers of norm 1 are ± 1. Show that this property does not hold for the algebraic integers in $\mathbb{Q}(\sqrt{-3})$.

3.4 Some Arithmetic in Dedekind Domains

Unique factorization of ideals in a Dedekind domain permits calculations that are analogous to familiar manipulations involving ordinary integers. In this section, we illustrate some of the ideas.

Let $P_1, \ldots, P_n$ be distinct nonzero prime ideals of the Dedekind domain R, and let $J = P_1 \cdots P_n$. Let Q_i be the product of the P_j with P_i omitted, that is,

$$Q_i = P_1 \cdots P_{i-1} P_{i+1} \cdots P_n.$$

(If $n = 1$, we take $Q_1 = R$.) If I is any nonzero ideal of R, then by unique factorization, $IQ_i \supset IJ$. For each $i = 1, \ldots, n$, choose an element a_i belonging to IQ_i but not to IJ, and let $a = \sum_{i=1}^{n} a_i$.

3.4.1 Lemma

The element a belongs to I, but for each i, $a \notin IP_i$. (In particular, $a \neq 0$.)

Proof. Since each a_i belongs to $IQ_i \subseteq I$, we have $a \in I$. Now a_i cannot belong to IP_i, for if so, $a_i \in IP_i \cap IQ_i$, which is the least common multiple of IP_i and IQ_i [see (3.3.6)]. But by definition of Q_i, the least common multiple is simply IJ, which contradicts the choice of a_i. We break up the sum defining a as follows:

$$a = (a_1 + \cdots + a_{i-1}) + a_i + (a_{i+1} + \cdots + a_n). \tag{1}$$

If $j \neq i$, then $a_j \in IQ_j \subseteq IP_i$, so the first and third terms of the right side of (1) belong to IP_i. Since $a_i \notin IP_i$, as found above, we have $a \notin IP_i$. ♣

In (3.3.7), we found that any nonzero ideal is a factor of a principal ideal. We can sharpen this result as follows.

3.4.2   Proposition

Let I be a nonzero ideal of the Dedekind domain R. Then there is a nonzero ideal I' such that II' is a principal ideal (a). Moreover, if J is an arbitrary nonzero ideal of R, then I' can be chosen to be relatively prime to J.

Proof. Let $P_1, \ldots, P_n$ be the distinct prime divisors of J, and choose a as in (3.4.1). Then $a \in I$, so $(a) \subseteq I$. Since divides means contains [see (3.3.5)], I divides (a), so $(a) = II'$ for some nonzero ideal I'. If I' is divisible by P_i, then $I' = P_i I_0$ for some nonzero ideal I_0, and $(a) = IP_i I_0$. Consequently, $a \in IP_i$, contradicting (3.4.1). ♣

3.4.3   Corollary

A Dedekind domain with only finitely many prime ideals is a PID.

Proof. Let J be the product of all the nonzero prime ideals. If I is any nonzero ideal, then by (3.4.2) there is a nonzero ideal I' such that II' is a principal ideal (a), with I' relatively prime to J. But then the set of prime factors of I' is empty, so $I' = R$. Thus $(a) = II' = IR = I$. ♣

The next result reinforces the idea that a Dedekind domain is not too far away from a principal ideal domain.

3.4.4   Corollary

Let I be a nonzero ideal of the Dedekind domain R, and let a be any nonzero element of I. Then I can be generated by two elements, one of which is a.

Proof. Since $a \in I$, we have $(a) \subseteq I$, so I divides (a), say $(a) = IJ$. By (3.4.2), there is a nonzero ideal I' such that II' is a principal ideal (b) and I' is relatively prime to J. If gcd stands for greatest common divisor, then the ideal generated by a and b is

$$\gcd((a),(b)) = \gcd(IJ, II') = I$$

because $\gcd(J, I') = (1)$. ♣

3.4.5   The Ideal Class Group

Let $I(R)$ be the group of nonzero fractional ideals of a Dedekind domain R. If $P(R)$ is the subset of $I(R)$ consisting of all nonzero *principal fractional ideals* $Rx, x \in K$, then $P(R)$ is a subgroup of $I(R)$. To see this, note that $(Rx)(Ry)^{-1} = (Rx)(Ry^{-1}) = Rxy^{-1}$, which belongs to $P(R)$. The quotient group $C(R) = I(R)/P(R)$ is called the *ideal class group* of R. Since R is commutative, $C(R)$ is abelian, and we will show later that in the number field case, $C(R)$ is finite.

Let us verify that $C(R)$ is trivial if and only if R is a PID. If $C(R)$ is trivial, then every integral ideal I of R is a principal fractional ideal $Rx, x \in K$. But $I \subseteq R$, so $x = 1x$ must belong to R, proving that R is a PID. Conversely, if R is a PID and I is a nonzero fractional ideal, then $rI \subseteq R$ for some nonzero $r \in R$. By hypothesis, the integral ideal rI must be principal, so $rI = Ra$ for some $a \in R$. Thus $I = R(a/r)$ with $a/r \in K$, and we conclude that every nonzero fractional ideal of R is a principal fractional ideal.

Problems For Section 3.4

We will now go through the factorization of an ideal in a number field. In the next chapter, we will begin to develop the necessary background, but some of the manipulations are accessible to us now. By (2.3.11), the ring B of algebraic integers of the number field $\mathbb{Q}(\sqrt{-5})$ is $\mathbb{Z}[\sqrt{-5}]$. (Note that $-5 \equiv 3 \mod 4$.) If we wish to factor the ideal $(2) = 2B$ of B, the idea is to factor $x^2 + 5 \mod 2$, and the result is $x^2 + 5 \equiv (x+1)^2 \mod 2$. Identifying x with $\sqrt{-5}$, we form the ideal $P_2 = (2, 1 + \sqrt{-5})$, which turns out to be prime. The desired factorization is $(2) = P_2^2$. This technique works if $B = \mathbb{Z}[\alpha]$, where the number field L is $\mathbb{Q}(\sqrt{\alpha})$.

1. Show that $1 - \sqrt{-5} \in P_2$, and conclude that $6 \in P_2^2$.
2. Show that $2 \in P_2^2$, hence $(2) \subseteq P_2^2$.
3. Expand $P_2^2 = (2, 1 + \sqrt{-5})(2, 1 + \sqrt{-5})$, and conclude that $P_2^2 \subseteq (2)$.
4. Following the technique suggested in the above problems, factor $x^2 + 5 \mod 3$, and conjecture that the prime factorization of (3) in the ring of algebraic integers of $\mathbb{Q}(\sqrt{-5})$ is $(3) = P_3 P_3'$ for appropriate P_3 and P_3'.
5. With P_3 and P_3' as found in Problem 4, verify that $(3) = P_3 P_3'$.

Chapter 4

Factoring of Prime Ideals in Extensions

4.1 Lifting of Prime Ideals

Recall the basic $AKLB$ setup: A is a Dedekind domain with fraction field K, L is a finite, separable extension of K of degree n, and B is the integral closure of A in L. If $A = \mathbb{Z}$, then $K = \mathbb{Q}$, L is a number field, and B is the ring of algebraic integers of L.

4.1.1 Definitions and Comments

Let P be a nonzero prime ideal of A. The *lifting* (also called the *extension*) of P to B is the ideal PB. Although PB need not be a prime ideal of B, we can use the fact that B is a Dedekind domain [see (3.1.3)] and the unique factorization theorem (3.3.1) to write

$$PB = \prod_{i=1}^{g} P_i^{e_i}$$

where the P_i are distinct prime ideals of B and the e_i are positive integers [see (3.3.2)].

On the other hand, we can start with a nonzero prime ideal Q of B and form a prime ideal of A via

$$P = Q \cap A.$$

We say that Q *lies over* P, or that P is the *contraction* of Q to A.

Now suppose that we start with a nonzero prime ideal P of A and lift it to B. We will show that the prime ideals $P_1, \ldots, P_g$ that appear in the prime factorization of PB are precisely the prime ideals of B that lie over P.

4.1.2 Proposition

Let Q be a nonzero prime ideal of B. Then Q appears in the prime factorization of PB if and only if $Q \cap A = P$.

Proof. If $Q \cap A = P$, then $P \subseteq Q$, hence $PB \subseteq Q$ because Q is an ideal. By (3.3.5), Q divides PB. Conversely, assume that Q divides, hence contains, PB. Then

$$P = P \cap A \subseteq PB \cap A \subseteq Q \cap A.$$

But in a Dedekind domain, every nonzero prime ideal is maximal, so $P = Q \cap A$. ♣

4.1.3   Ramification and Relative Degree

If we lift P to B and factor PB as $\prod_{i=1}^{g} P_i^{e_i}$, the positive integer e_i is called the *ramification index* of P_i over P (or over A). We say that P *ramifies* in B (or in L) if $e_i > 1$ for at least one i. We will prove in a moment that B/P_i is a finite extension of the field A/P. The degree f_i of this extension is called the *relative degree* (or the *residue class degree*, or the *inertial degree*) of P_i over P (or over A).

4.1.4   Proposition

We can identify A/P with a subfield of B/P_i, and B/P_i is a finite extension of A/P.

Proof. The map from A/P to B/P_i given by $a + P \to a + P_i$ is well-defined and injective, because $P = P_i \cap A$, and it is a homomorphism by direct verification. By (3.1.2), B is a finitely generated A-module, hence B/P_i is a finitely generated A/P-module, that is, a finite-dimensional vector space over A/P. ♣

4.1.5   Remarks

The same argument, with P_i replaced by PB, shows that B/PB is a finitely generated A/P-algebra, in particular, a finite-dimensional vector space over A/P. We will denote the dimension of this vector space by $[B/PB : A/P]$.

The numbers e_i and f_i are connected by an important identity, which does not seem to have a name in the literature. We will therefore christen it as follows.

4.1.6   Ram-Rel Identity

$$\sum_{i=1}^{g} e_i f_i = [B/PB : A/P] = n.$$

Proof. To prove the first equality, consider the chain of ideals

$$B \supseteq P_1 \supseteq P_1^2 \supseteq \cdots \supseteq P_1^{e_1}$$
$$\supseteq P_1^{e_1} P_2 \supseteq P_1^{e_1} P_2^2 \supseteq \cdots \supseteq P_1^{e_1} P_2^{e_2}$$
$$\supseteq \cdots \supseteq P_1^{e_1} \cdots P_g^{e_g} = PB.$$

By unique factorization, there can be no ideals between consecutive terms in the sequence. (Any such ideal would contain, hence divide, PB.) Thus the quotient $\beta/\beta P_i$ of any two consecutive terms is a one-dimensional vector space over B/P_i, as there are no nontrivial proper subspaces. (It is a vector space over this field because it is annihilated by P_i.) But, with notation as in (4.1.5), $[B/P_i : A/P] = f_i$, so $[\beta/\beta P_i : A/P] = f_i$. For each i, we have exactly e_i consecutive quotients, each of dimension f_i over A/P. Consequently, $[B/PB : A/P] = \sum_{i=1}^{g} e_i f_i$, as claimed.

To prove the second equality, we first assume that B is a free A-module of rank n. By (2.3.8), this covers the case where A is a PID, in particular, when L is a number field. If $x_1, \ldots, x_n$ is a basis for B over A, we can reduce mod PB to produce a basis for B/PB over A/P, and the result follows. Explicitly, suppose $\sum_{i=1}^{n}(a_i+P)(x_i+PB) = 0$ in B/PB. Then $\sum_{i=1}^{n} a_i x_i$ belongs to PB, hence can be written as $\sum_j b_j y_j$ with $b_j \in B, y_j \in P$. Since $b_j = \sum_k c_{jk} x_k$ with $c_{jk} \in A$, we have $a_k = \sum_j c_{jk} y_j \in P$ for all k.

The general case is handled by localization. Let $S = A \backslash P$, $A' = S^{-1}A, B' = S^{-1}B$. By (1.2.6), (1.2.9), and the Dedekind property (every nonzero prime ideal of A is maximal), it follows that A' has exactly one nonzero prime ideal, namely $P' = PA'$. Moreover, P' is principal, so A' is a *discrete valuation ring*, that is, a local PID that is not a field. [By unique factorization, we can choose an element $a \in P' \backslash (P')^2$, so $(a) \subseteq P'$ but $(a) \not\subseteq (P')^2$. Since the only nonzero ideals of A' are powers of P' (unique factorization again), we have $(a) = P'$.] Now B is the integral closure of A in L, so B' is the integral closure of A' in $S^{-1}L = L$. [The idea is that we can go back and forth between an equation of integral dependence for $b \in B$ and an equation of integral dependence for $b/s \in B'$ either by introducing or clearing denominators.] We have now reduced to the PID case already analyzed, and $[B'/PB' : A'/PA'] = n$.

Now $PB = \prod_{i=1}^{g} P_i^{e_i}$, and P_i is a nonzero prime ideal of B not meeting S. [If $y \in P_i \cap S$, then $y \in P_i \cap A = P$ by (4.1.2). Thus $y \in P \cap S$, a contradiction.] By the basic correspondence (1.2.6), we have the factorization $PB' = \prod_{i=1}^{g} (P_i B')^{e_i}$. By the PID case,

$$n = [B'/PB' : A'/PA'] = \sum_{i=1}^{g} e_i [B'/P_i B' : A'/PA'].$$

We are finished if we can show that $B'/P_i B' \cong B/P_i$ and $A'/PA' \cong A/P$. The statement of the appropriate lemma, and the proof in outline form, are given in the exercises. ♣

Problems For Section 4.1

We will fill in the gap at the end of the proof of the ram-rel identity. Let S be a multiplicative subset of the integral domain A, and let $\mathcal{M}$ be a maximal ideal of A disjoint from S. Consider the composite map $A \to S^{-1}A \to S^{-1}A/\mathcal{M}S^{-1}A$, where the first map is given by $a \to a/1$ and the second by $a/s \to (a/s) + \mathcal{M}S^{-1}A$.

1. Show that the kernel of the map is $\mathcal{M}$, so by the factor theorem, we have a monomorphism $h : A/\mathcal{M} \to S^{-1}A/\mathcal{M}S^{-1}A$.
2. Let $a/s \in S^{-1}A$. Show that for some $b \in A$ we have $bs \equiv 1 \mod \mathcal{M}$.
3. Show that $(a/s) + \mathcal{M}S^{-1}A = h(ab)$, so h is surjective and therefore an isomorphism.

Consequently, $S^{-1}A/\mathcal{M}S^{-1}A \cong A/\mathcal{M}$, which is the result we need.

4.2 Norms of Ideals

4.2.1 Definitions and Comments

We are familiar with the norm of an element of a field, and we are going to extend the idea to ideals. We assume the $AKLB$ setup with $A = \mathbb{Z}$, so that B is a *number ring*, that is, the ring of algebraic integers of a number field L. If I is a nonzero ideal of B, we define the *norm* of I by $N(I) = |B/I|$. We will show that the norm is finite, so if P is a nonzero prime ideal of B, then B/P is a finite field. Also, N has a multiplicative property analogous to the formula $N(xy) = N(x)N(y)$ for elements. [See (2.1.3), equation (2).]

4.2.2 Proposition

Let b be any nonzero element of the ideal I of B, and let $m = N_{L/\mathbb{Q}}(b) \in \mathbb{Z}$. Then $m \in I$ and $|B/mB| = m^n$, where $n = [L : \mathbb{Q}]$.

Proof. By (2.1.6), $m = bc$ where c is a product of conjugates of b. But a conjugate of an algebraic integer is an algebraic integer. (If a monomorphism is applied to an equation of integral dependence, the result is an equation of integral dependence.) Thus $c \in B$, and since $b \in I$, we have $m \in I$. Now by (2.3.9), B is the direct sum of n copies of $\mathbb{Z}$, hence by the first isomorphism theorem, B/mB is the direct sum of n copies of $\mathbb{Z}/m\mathbb{Z}$. Consequently, $|B/mB| = m^n$. ♣

4.2.3 Corollary

If I is any nonzero ideal of B, then $N(I)$ is finite. In fact, if m is as in (4.2.2), then $N(I)$ divides m^n.

Proof. Observe that $(m) \subseteq I$, hence

$$\frac{B/(m)}{B/I} \cong I/(m). \quad ♣$$

4.2.4 Corollary

Every nonzero ideal I of B is a free abelian group of rank n.

Proof. By the simultaneous basis theorem, we may represent B as the direct sum of n copies of $\mathbb{Z}$, and I as the direct sum of $a_1\mathbb{Z}, \dots, a_r\mathbb{Z}$, where $r \leq n$ and the a_i are positive integers such that a_i divides a_{i+1} for all i. Thus B/I is the direct sum of r cyclic groups (whose orders are $a_1, \dots, a_r$) and $n - r$ copies of $\mathbb{Z}$. If $r < n$, then at least one copy of $\mathbb{Z}$ appears, and $|B/I|$ cannot be finite. ♣

4.2.5 Computation of the Norm

Suppose that $\{x_1, \dots, x_n\}$ is a $\mathbb{Z}$-basis for B, and $\{z_1, \dots, z_n\}$ is a basis for I. Each z_i is a linear combination of the x_i with integer coefficients, in matrix form $z = Cx$. We claim that the norm of I is the absolute value of the determinant of C. To verify this, first look at the special case $x_i = y_i$ and $z_i = a_i y_i$, as in the proof of (4.2.4). Then C is a diagonal matrix with entries a_i, and the result follows. But the special case implies the general result, because any matrix corresponding to a change of basis of B or I is unimodular, in other words, has integer entries and determinant ± 1. [See (2.3.9) and (2.3.10).]

Now with $z = Cx$ as above, the discriminant of x is the field discriminant d, and the discriminant of z is $D(z) = (\det C)^2 d$ by (2.3.2). We have just seen that $N(I) = |\det C|$, so we have the following formula for computing the norm of an ideal I. If z is a $\mathbb{Z}$-basis for I, then

$$N(I) = \left| \frac{D(z)}{d} \right|^{1/2}.$$

There is a natural relation between the norm of a principal ideal and the norm of the corresponding element.

4.2.6 Proposition

If $I = (a)$ with $a \neq 0$, then $N(I) = |N_{L/\mathbb{Q}}(a)|$.

Proof. If x is a $\mathbb{Z}$-basis for B, then ax is a $\mathbb{Z}$-basis for I. By (2.3.3), $D(ax)$ is the square of the determinant whose ij entry is $\sigma_i(ax_j) = \sigma_i(a)\sigma_i(x_j)$. By (4.2.5), the norm of I is $|\sigma_1(a) \cdots \sigma_n(a)| = |N_{L/\mathbb{Q}}(a)|$. ♣

In the proof of (4.2.6), we cannot invoke (2.3.2) to get $D(ax_1, \ldots, ax_n) = (a^n)^2 D(x_1, \ldots, x_n)$, because we need not have $a \in \mathbb{Q}$.

We now establish the multiplicative property of ideal norms.

4.2.7 Theorem

If I and J are nonzero ideals of B, then $N(IJ) = N(I)N(J)$.

Proof. By unique factorization, we may assume without loss of generality that J is a prime ideal P. By the third isomorphism theorem, $|B/IP| = |B/I| \, |I/IP|$, so we must show that $|I/IP|$ is the norm of P, that is, $|B/P|$. But this has already been done in the first part of the proof of (4.1.6). ♣

4.2.8 Corollary

Let I be a nonzero ideal of B. If $N(I)$ is prime, then I is a prime ideal.

Proof. Suppose I is the product of two ideals I_1 and I_2. By (4.2.7), $N(I) = N(I_1)N(I_2)$, so by hypothesis, $N(I_1) = 1$ or $N(I_2) = 1$. Thus either I_1 or I_2 is the identity element of the ideal group, namely B. Therefore, the prime factorization of I is I itself, in other words, I is a prime ideal. ♣

4.2.9 Proposition

$N(I) \in I$, in other words, I divides $N(I)$. [More precisely, I divides the principal ideal generated by $N(I)$.]

Proof. Let $N(I) = |B/I| = r$. If $x \in B$, then $r(x + I)$ is 0 in B/I, because the order of any element of a group divides the order of the group. Thus $rx \in I$, and in particular we may take $x = 1$ to conclude that $r \in I$. ♣

4.2.10   Corollary

If I is a nonzero prime ideal of B, then I divides (equivalently, contains) exactly one rational prime p.

Proof. By (4.2.9), I divides $N(I) = p_1^{m_1} \cdots p_t^{m_t}$, so I divides some p_i. But if I divides two distinct primes p and q, then there exist integers u and v such that $up + vq = 1$. Thus I divides 1, so $I = B$, a contradiction. Therefore I divides exactly one p. ♣

4.2.11   The Norm of a Prime Ideal

If we can compute the norm of every nonzero prime ideal P, then by multiplicativity, we can calculate the norm of any nonzero ideal. Let p be the unique rational prime in P, and recall from (4.1.3) that the relative degree of P over p is $f(P) = [B/P : \mathbb{Z}/p\mathbb{Z}]$. Therefore

$$N(P) = |B/P| = p^{f(P)}.$$

Note that by (4.2.6), the norm of the principal ideal (p) is $|N(p)| = p^n$, so $N(P) = p^m$ for some $m \leq n$. This conclusion also follows from the above formula $N(P) = p^{f(P)}$ and the ram-rel identity (4.1.6).

Here are two other useful finiteness results.

4.2.12   Proposition

A rational integer m can belong to only finitely many ideals of B.

Proof. We have $m \in I$ iff I divides (m), and by unique factorization, (m) has only finitely many divisors. ♣

4.2.13   Corollary

Only finitely many ideals can have a given norm.

Proof. If $N(I) = m$, then by (4.2.9), $m \in I$, and the result follows from (4.2.12). ♣

Problems For Section 4.2

This problem set will give the proof that a rational prime p ramifies in the number field L if and only if p divides the field discriminant $d = d_{L/\mathbb{Q}}$.

1. Let $(p) = pB$ have prime factorization $\prod_i P_i^{e_i}$. Show that p ramifies if and only if the ring $B/(p)$ has nonzero nilpotent elements.

Now as in (2.1.1), represent elements of B by matrices with respect to an integral basis $\omega_1, \ldots, \omega_n$ of B. Reduction of the entries mod p gives matrices representing elements of $B/(p)$.

2. Show that a nilpotent element (or matrix) has zero trace.

Suppose that $A(\beta)$, the matrix representing the element β, is nilpotent mod p. Then $A(\beta\omega_i)$ will be nilpotent mod p for all i, because $\beta\omega_i$ is nilpotent mod p.

3. By expressing β in terms of the ω_i and computing the trace of $A(\beta\omega_j)$, show that if β is nilpotent mod p and $\beta \notin (p)$, then $d \equiv 0 \mod p$, hence p divides d.

Now assume that p does not ramify.

4. Show that $B/(p)$ is isomorphic to a finite product of finite fields F_i of characteristic p.

Let $\pi_i : B \to B/(p) \to F_i$ be the composition of the canonical map from B onto $B/(p)$ and the projection from $B/(p)$ onto F_i.

5. Show that the trace form $T_i(x, y) = T_{F_i/\mathbb{F}_p}(\pi_i(x)\pi_i(y))$ is nondegenerate, and conclude that $\sum_i T_i$ is also nondegenerate.

We have $d = \det T(\omega_i\omega_j)$, in other words, the determinant of the matrix of the bilinear form $T(x, y)$ on B, with respect to the basis $\{\omega_1, \dots, \omega_n\}$. Reducing the matrix entries mod p, we get the matrix of the reduced bilinear form T_0 on the $\mathbb{F}_p$-vector space $B/(p)$.

6. Show that T_0 coincides with $\sum_i T_i$, hence T_0 is nondegenerate. Therefore $d \not\equiv 0 \mod p$, so p does not divide d.

As a corollary, it follows that only finitely many primes can ramify in L.

4.3   A Practical Factorization Theorem

The following result, usually credited to Kummer but sometimes attributed to Dedekind, allows, under certain conditions, an efficient factorization of a rational prime in a number field.

4.3.1   Theorem

Let L be a number field of degree n over $\mathbb{Q}$, and assume that the ring B of algebraic integers of L is $\mathbb{Z}[\theta]$ for some $\theta \in B$. Thus $1, \theta, \theta^2, \dots, \theta^{n-1}$ form an integral basis of B. Let p be a rational prime, and let f be the minimal polynomial of θ over $\mathbb{Q}$. Reduce the coefficients of f modulo p to obtain $\overline{f} \in \mathbb{Z}[X]$. Suppose that the factorization of $\overline{f}$ into irreducible polynomials over $\mathbb{F}_p$ is given by

$$\overline{f} = h_1^{e_1} \cdots h_r^{e_r}.$$

Let f_i be any polynomial in $\mathbb{Z}[X]$ whose reduction mod p is h_i. Then the ideal

$$P_i = (p, f_i(\theta))$$

is prime, and the prime factorization of (p) in B is

$$(p) = P_1^{e_1} \cdots P_r^{e_r}.$$

Proof. Adjoin a root θ_i of h_i to produce the field $\mathbb{F}_p[\theta_i] \cong \mathbb{F}_p[X]/h_i(X)$. The assignment $\theta \to \theta_i$ extends by linearity (and reduction of coefficients mod p) to an epimorphism $\lambda_i : \mathbb{Z}[\theta] \to \mathbb{F}_p[\theta_i]$. Since $\mathbb{F}_p[\theta_i]$ is a field, the kernel of λ_i is a maximal, hence prime, ideal of $\mathbb{Z}[\theta] = B$. Since λ_i maps $f_i(\theta)$ to $h_i(\theta_i) = 0$ and also maps p to 0, it follows that $P_i \subseteq \ker \lambda_i$. We claim that $P_i = \ker \lambda_i$. To prove this, assume $g(\theta) \in \ker \lambda_i$. With a subscript 0 indicating reduction of coefficients mod p, we have $g_0(\theta_i) = 0$, hence h_i, the minimal polynomial of θ_i, divides g_0. If $g_0 = q_0 h_i$, then $g - q f_i \equiv 0 \mod p$. Therefore

$$g(\theta) = [g(\theta) - q(\theta) f_i(\theta)] + q(\theta) f_i(\theta)$$

so $g(\theta)$ is the sum of an element of (p) and an element of $(f_i(\theta))$. Thus $\ker \lambda_i \subseteq P_i$, so $P_i = \ker \lambda_i$, a prime ideal.

We now show that (p) divides $P_1^{e_1} \cdots P_r^{e_r}$. We use the identity $(I+I_1)(I+I_2) \subseteq I+I_1 I_2$, where I, I_1 and I_2 are ideals. We begin with $P_1 = (p) + (f_1(\theta))$, and compute

$$P_1^2 \subseteq (p) + (f_1(\theta))^2, \ldots, P_1^{e_1} \cdots P_r^{e_r} \subseteq (p) + (f_1(\theta))^{e_1} \cdots (f_r(\theta))^{e_r}.$$

But the product of the $f_i(\theta)^{e_i}$ coincides mod p with $\prod_{i=1}^r h_i(\theta) = \overline{f}(\theta) = 0$. We conclude that $\prod_{i=1}^r P_i^{e_i} \subseteq (p)$, as asserted.

We now know that $(p) = P_1^{k_1} \cdots P_r^{k_r}$ with $0 \leq k_i \leq e_i$. (Actually, $k_i > 0$ since $p \in \ker \lambda_i = P_i$, so P_i divides (p). But we will not need this refinement.) By hypothesis, $B/P_i = \mathbb{Z}[\theta]/P_i$, which is isomorphic to $\mathbb{F}_p[\theta_i]$, as observed at the beginning of the proof. Thus the norm of P_i is $|\mathbb{F}_p[\theta_i]| = p^{d_i}$, where d_i is the degree of h_i. By (4.2.6), (4.2.7) and equation (3) of (2.1.3),

$$p^n = N((p)) = \prod_{i=1}^r N(P_i)^{k_i} = \prod_{i=1}^r p^{d_i k_i}$$

hence $n = d_1 k_1 + \cdots + d_r k_r$. But n is the degree of the monic polynomial f, which is the same as $\deg \overline{f} = d_1 e_1 + \cdots + d_r e_r$. Since $k_i \leq e_i$ for every i, we have $k_i = e_i$ for all i, and the result follows. ♣

4.3.2   Prime Factorization in Quadratic Fields

We consider $L = \mathbb{Q}(\sqrt{m})$, where m is a square-free integer, and factor the ideal (p) in the ring B of algebraic integers of L. By the ram-rel identity (4.1.6), there will be three cases:

(1) $g = 2, e_1 = e_2 = f_1 = f_2 = 1$. Then (p) is the product of two distinct prime ideals P_1 and P_2, and we say that p *splits* in L.

(2) $g = 1, e_1 = 1, f_1 = 2$. Then (p) is a prime ideal of B, and we say that p *remains prime* in L or that p is *inert*.

(3) $g = 1, e_1 = 2, f_1 = 1$. Then $(p) = P_1^2$ for some prime ideal P_1, and we say that p *ramifies* in L.

We will examine all possibilities systematically.

(a) Assume p is an odd prime not dividing m. Then p does not divide the discriminant, so p does not ramify.

(a1) If m is a quadratic residue mod p, then p splits. Say $m \equiv n^2 \mod p$. Then $x^2 - m$ factors mod p as $(x+n)(x-n)$, so $(p) = (p, n + \sqrt{m})\,(p, n - \sqrt{m})$.

(a2) If m is not a quadratic residue mod p, then $x^2 - m$ cannot be the product of two linear factors, hence $x^2 - m$ is irreducible mod p and p remains prime.

(b) Let p be any prime dividing m. Then p divides the discriminant, hence p ramifies. Since $x^2 - m \equiv x^2 = xx \mod p$, we have $(p) = (p, \sqrt{m})^2$.

This takes care of all odd primes, and also $p = 2$ with m even.

(c) Assume $p = 2$, m odd.

(c1) Let $m \equiv 3 \mod 4$. Then 2 divides the discriminant $D = 4m$, so 2 ramifies. We have $x^2 - m \equiv (x+1)^2 \mod 2$, so $(2) = (2, 1 + \sqrt{m})^2$.

(c2) Let $m \equiv 1 \mod 8$, hence $m \equiv 1 \mod 4$. An integral basis is $\{1, (1 + \sqrt{m})/2\}$, and the discriminant is $D = m$. Thus 2 does not divide D, so 2 does not ramify. We claim that $(2) = (2, (1 + \sqrt{m})/2) (2, (1 - \sqrt{m})/2)$. To verify this note that the right side is $(2, 1 - \sqrt{m}, 1 + \sqrt{m}, (1 - m)/4)$. This coincides with (2) because $(1 - m)/4$ is an even integer and $1 - \sqrt{m} + 1 + \sqrt{m} = 2$.

If $m \equiv 3$ or 7 mod 8, then $m \equiv 3 \mod 4$, so there is only one remaining case.

(c3) Let $m \equiv 5 \mod 8$, hence $m \equiv 1 \mod 4$, so $D = m$ and 2 does not ramify. Consider $f(x) = x^2 - x + (1 - m)/4$ over B/P, where P is any prime ideal lying over (2). The roots of f are $(1 \pm \sqrt{m})/2$, so f has a root in B, hence in B/P. But there is no root in $\mathbb{F}_2$, because $(1 - m)/4 \equiv 1 \mod 2$. Thus B/P and $\mathbb{F}_2$ cannot be isomorphic. If (2) factors as $Q_1 Q_2$, then the norm of (2) is 4, so Q_1 and Q_2 have norm 2, so the B/Q_i are isomorphic to $\mathbb{F}_2$, which contradicts the argument just given. Therefore 2 remains prime.

You probably noticed something suspicious in cases (a) and (b). In order to apply (4.3.1), 1 and $\sqrt{m}$ must form an integral basis, so $m \not\equiv 1 \mod 4$, as in (2.3.11). But we can repair the damage. In (a1), verify directly that the factorization of (p) is as given. The key point is that the ideal $(p, n + \sqrt{m}) (p, n - \sqrt{m})$ contains $p(n + \sqrt{m} + n - \sqrt{m}) = 2np$, and if p divides n, then p divides $(m - n^2) + n^2 = m$, contradicting the assumption of case (a). Thus the greatest common divisor of p^2 and $2np$ is p, so p belongs to the ideal. Since every generator of the ideal is a multiple of p, the result follows. In (a2), suppose $(p) = Q_1 Q_2$. Since the norm of p is p^2, each Q_i has norm p, so B/Q_i must be isomorphic to $\mathbb{F}_p$. But $\sqrt{m} \in B$, so m has a square root in B/Q_i [see (4.1.4)]. But case (a2) assumes that there is no square root of m in $\mathbb{F}_p$, a contradiction. Finally, case (b) is similar to case (a1). We have $p|m$, but p^2 does not divide the square-free integer m, so the greatest common divisor of p^2 and m is p.

Problems For Section 4.3

1. In the exercises for Section 3.4, we factored (2) and (3) in the ring B of algebraic integers of $L = \mathbb{Q}(\sqrt{-5})$, using ad hoc techniques. Using the results of this section, derive the results rigorously.

2. Continuing Problem 1, factor (5), (7) and (11).

3. Let $L = \mathbb{Q}(\sqrt[3]{2})$, and assume as known that the ring of algebraic integers is $B = \mathbb{Z}[\sqrt[3]{2}]$. Find the prime factorization of (5).

Chapter 5

The Ideal Class Group

We will use Minkowski theory, which belongs to the general area of geometry of numbers, to gain insight into the ideal class group of a number field. We have already mentioned the ideal class group briefly in (3.4.5); it measures how close a Dedekind domain is to a principal ideal domain.

5.1 Lattices

5.1.1 Definitions and Comments

Let $e_1, \ldots, e_n \in \mathbb{R}^n$, with the e_i linearly independent over $\mathbb{R}$. Thus the e_i form a basis for $\mathbb{R}^n$ as a vector space over $\mathbb{R}$. The e_i also form a basis for a free $\mathbb{Z}$-module of rank n, namely

$$H = \mathbb{Z}e_1 + \cdots + \mathbb{Z}e_n.$$

A set H constructed in this way is said to be a *lattice* in $\mathbb{R}^n$. The *fundamental domain* of H is given by

$$T = \{x \in \mathbb{R}^n : x = \sum_{i=1}^{n} a_i e_i, \ 0 \le a_i < 1\}.$$

In the most familiar case, e_1 and e_2 are linearly independent vectors in the plane, and T is the parallelogram generated by the e_i. In general, every point of $\mathbb{R}^n$ is congruent modulo H to a unique point of T, so $\mathbb{R}^n$ is the disjoint union of the sets $h + T$, $h \in H$. If μ is Lebesgue measure, then the volume $\mu(T)$ of the fundamental domain T will be denoted by $v(H)$. If we generate H using a different $\mathbb{Z}$-basis, the volume of the fundamental domain is unchanged. (The change of variables matrix between $\mathbb{Z}$-bases is unimodular, hence has determinant ± 1. The result follows from the change of variables formula for multiple integrals.)

5.1.2   Lemma

Let S be a Lebesgue measurable subset of $\mathbb{R}^n$ with $\mu(S) > v(H)$. Then there exist distinct points $x, y \in S$ such that $x - y \in H$.

Proof. As we observed in (5.1.1), the sets $h + T, h \in H$, are (pairwise) disjoint and cover $\mathbb{R}^n$. Thus the sets $S \cap (h + T), h \in H$, are disjoint and cover S. Consequently,

$$\mu(S) = \sum_{h \in H} \mu(S \cap (h + T)).$$

By translation-invariance of Lebesgue measure, $\mu(S \cap (h + T)) = \mu((-h + S) \cap T)$. Now if $S \cap (h_1 + T)$ and $S \cap (h_2 + T)$ are disjoint, it does not follow that $(-h_1 + S) \cap T$ and $(-h_2 + S) \cap T$ are disjoint, as we are not subtracting the same vector from each set. In fact, if the sets $(-h + S) \cap T, h \in H$, were disjoint, we would reach a contradiction via

$$v(H) = \mu(T) \geq \sum_{h \in H} \mu((-h + S) \cap T) = \mu(S).$$

Thus there are distinct elements $h_1, h_2 \in H$ such that $(-h_1 + S) \cap (-h_2 + S) \cap T \neq \emptyset$. Choose (necessarily distinct) $x, y \in S$ such that $-h_1 + x = -h_2 + y$. Then $x - y = h_1 - h_2 \in H$, as desired. ♣

5.1.3   Minkowski's Convex Body Theorem

Let H be a lattice in $\mathbb{R}^n$, and assume that S is a Lebesgue measurable subset of $\mathbb{R}^n$ that is symmetric about the origin and convex. If

(a) $\mu(S) > 2^n v(H)$, or

(b) $\mu(S) \geq 2^n v(H)$ and S is compact,

then $S \cap (H \setminus \{0\}) \neq \emptyset$.

Proof.

(a) Let $S' = \frac{1}{2}S$. Then $\mu(S') = 2^{-n}\mu(S) > v(H)$ by hypothesis, so by (5.1.2), there exist distinct elements $y, z \in S'$ such that $y - z \in H$. But $y - z = \frac{1}{2}(2y + (-2z))$, a convex combination of $2y$ and $-2z$. But $y \in S' \Rightarrow 2y \in S$, and $z \in S' \Rightarrow 2z \in S \Rightarrow -2z \in S$ by symmetry about the origin. Thus $y - z \in S$ and since y and z are distinct, $y - z \in H \setminus \{0\}$.

(b) We apply (a) to $(1 + 1/m)S, m = 1, 2, \ldots$. Since S, hence $(1 + 1/m)S$, is a bounded set, it contains only finitely many points of the lattice H. Consequently, for every positive integer m, $S_m = (1 + 1/m)S \cap (H \setminus \{0\})$ is a nonempty finite, hence compact, subset of $\mathbb{R}^n$. Since $S_{m+1} \subseteq S_m$ for all m, the sets S_m form a nested sequence, and therefore $\cap_{m=1}^{\infty} S_m \neq \emptyset$. If $x \in \cap_{m=1}^{\infty} S_m$, then $x \in H \setminus \{0\}$ and $x/(1 + 1/m) \in S$ for every m. Since S is closed, we may let $m \to \infty$ to conclude that $x \in S$. ♣

5.1.4   Example

With $n = 2$, take $e_1 = (1, 0)$ and $e_2 = (0, 1)$. The fundamental domain is the unit square, closed at the bottom and on the left, and open at the top and on the right. Let S be the set of all $a_1 e_1 + a_2 e_2$ with $-1 < a_i < 1, i = 1, 2$. Then $\mu(S) = 4v(H)$, but S contains no nonzero lattice points. Thus compactness is a necessary hypothesis in part (b).

5.2   A Volume Calculation

We will use n-dimensional integration technique to derive a result that will be needed in the proof that the ideal class group is finite. We will work in $\mathbb{R}^n$, realized as the product of r_1 copies of $\mathbb{R}$ and r_2 copies of $\mathbb{C}$, where $r_1 + 2r_2 = n$. Our interest is in the set

$$B_t = \{(y_1, \ldots, y_{r_1}, z_1, \ldots, z_{r_2}) \in \mathbb{R}^{r_1} \times \mathbb{C}^{r_2} : \sum_{i=1}^{r_1} |y_i| + 2 \sum_{j=1}^{r_2} |z_j| \leq t\}, t \geq 0.$$

We will show that the volume of B_t is given by

$$V(r_1, r_2, t) = 2^{r_1} \left(\frac{\pi}{2}\right)^{r_2} \frac{t^n}{n!}.$$

The proof is by double induction on r_1 and r_2. If $r_1 = 1$ and $r_2 = 0$, hence $n = 1$, we are calculating the length of the interval $[-t, t]$, which is $2t$, as predicted. If $r_1 = 0$ and $r_2 = 1$, hence $n = 2$, we are calculating the area of $\{z_1 : 2|z_1| \leq t\}$, a disk of radius $t/2$. The result is $\pi t^2/4$, again as predicted. Now assume that the formula holds for r_1, r_2, and all t. Then $V(r_1 + 1, r_2, t)$ is the volume of the set described by

$$|y| + \sum_{i=1}^{r_1} |y_i| + 2 \sum_{j=1}^{r_2} |z_j| \leq t$$

or equivalently by

$$\sum_{i=1}^{r_1} |y_i| + 2 \sum_{j=1}^{r_2} |z_j| \leq t - |y|.$$

Now if $|y| > t$, then B_t is empty. For smaller values of $|y|$, suppose we change $|y|$ to $|y| + dy$. This creates a box in $(n + 1)$-space with dy as one of the dimensions. The volume of the box is $V(r_1, r_2, t - |y|)dy$. Thus

$$V(r_1 + 1, r_2, t) = \int_{-t}^{t} V(r_1, r_2, t - |y|)dy$$

which by the induction hypothesis is $2 \int_0^t 2^{r_1} (\pi/2)^{r_2} [(t - y)^n/n!] \, dy$. Evaluating the integral, we obtain $2^{r_1+1}(\pi/2)^{r_2} t^{n+1}/(n+1)!$, as desired.

Finally, $V(r_1, r_2 + 1, t)$ is the volume of the set described by

$$\sum_{i=1}^{r_1} |y_i| + 2 \sum_{j=1}^{r_2} |z_j| + 2|z| \leq t.$$

As above,

$$V(r_1, r_2 + 1, t) = \int_{|z| \leq t/2} V(r_1, r_2, t - 2|z|)d\mu(z)$$

where μ is Lebesgue measure on $\mathbb{C}$. In polar coordinates, the integral becomes

$$\int_{\theta=0}^{2\pi} \int_{r=0}^{t/2} 2^{r_1} \left(\frac{\pi}{2}\right)^{r_2} \frac{(t-2r)^n}{n!} \, r \, dr \, d\theta$$

which reduces to $2^{r_1}(\pi/2)^{r_2}(2\pi/n!)\int_{r=0}^{t/2}(t-2r)^n \, r \, dr$. We may write the integrand as $(t-2r)^n \, r \, dr = -rd(t-2r)^{n+1}/2(n+1)$. Integration by parts yields (for the moment ignoring the constant factors preceding the integral)

$$\int_0^{t/2} (t-2r)^{n+1} dr/2(n+1) = \left. \frac{-(t-2r)^{n+2}}{2(n+1)2(n+2)} \right|_0^{t/2} = \frac{t^{n+2}}{4(n+1)(n+2)}.$$

Therefore $V(r_1, r_2+1, t) = 2^{r_1}(\pi/2)^{r_2}(2\pi/n!)t^{n+2}/4(n+1)(n+2)$, which simplifies to $2^{r_1}(\pi/2)^{r_2+1}t^{n+2}/(n+2)!$, completing the induction. Note that $n+2$ (rather than $n+1$) is correct, because $r_1 + 2(r_2+1) = r_1 + 2r_2 + 2 = n + 2$.

5.3   The Canonical Embedding

5.3.1   Definitions and Comments

Let L be a number field of degree n over $\mathbb{Q}$, and let $\sigma_1, \ldots, \sigma_n$ be the $\mathbb{Q}$-monomorphisms of L into $\mathbb{C}$. If σ_i maps entirely into $\mathbb{R}$, we say that σ_i is a *real embedding*; otherwise it is a *complex embedding*. Since the complex conjugate of a $\mathbb{Q}$-monomorphism is also a $\mathbb{Q}$-monomorphism, we can renumber the σ_i so that the real embeddings are $\sigma_1, \ldots, \sigma_{r_1}$ and the complex embeddings are $\sigma_{r_1+1}, \ldots, \sigma_n$, with σ_{r_1+j} paired with its complex conjugate $\sigma_{r_1+r_2+j}$, $j = 1, \ldots, r_2$. Thus there are $2r_2$ complex embeddings, and $r_1 + 2r_2 = n$.

The *canonical embedding* $\sigma : L \to \mathbb{R}^{r_1} \times \mathbb{C}^{r_2} = \mathbb{R}^n$ is the injective ring homomorhism given by

$$\sigma(x) = (\sigma_1(x), \ldots, \sigma_{r_1+r_2}(x)).$$

5.3.2   Some Matrix Manipulations

Let $x_1, \ldots, x_n \in L$ be linearly dependent over $\mathbb{Z}$ (hence the x_i form a basis for L over $\mathbb{Q}$). Let C be the matrix whose k^{th} column ($k = 1, \ldots, n$) is

$$\sigma_1(x_k), \ldots, \sigma_{r_1}(x_k), \text{Re } \sigma_{r_1+1}(x_k), \text{Im } \sigma_{r_1+1}(x_k), \ldots, \text{Re } \sigma_{r_1+r_2}(x_k), \text{Im } \sigma_{r_1+r_2}(x_k).$$

The determinant of C looks something like a discriminant, and we can be more precise with the aid of elementary row operations. Suppose that

$$\begin{pmatrix} \sigma_j(x_k) \\ \overline{\sigma}_j(x_k) \end{pmatrix} = \begin{pmatrix} x+iy \\ x-iy \end{pmatrix}.$$

We are fixing j and allowing k to range from 1 to n, so we have two rows of an n by n matrix. Add the second row to the first, so that the entries on the right become $2x$

and $x - iy$. Then add $-1/2$ times row 1 to row 2, and the entries become $2x$ and $-iy$. Factoring out 2 and $-i$, we get

$$-2i \begin{pmatrix} x \\ y \end{pmatrix} = -2i \begin{pmatrix} \operatorname{Re} \sigma_j(x_k) \\ \operatorname{Im} \sigma_j(x_k) \end{pmatrix}.$$

Do this for each $j = 1, \ldots, r_2$. In the above calculation, $\overline{\sigma}_j$ appears immediately under σ_j, but in the original ordering they are separated by r_2, which introduces a factor of $(-1)^{r_2}$ when we calculate a determinant. To summarize, we have

$$\det C = (2i)^{-r_2} \det(\sigma_j(x_k))$$

Note that j and k range from 1 to n; no operations are needed for the first r_1 rows.

Now let M be the free $\mathbb{Z}$-module generated by the x_i, so that $\sigma(M)$ is a free $\mathbb{Z}$-module with basis $\sigma(x_i), i = 1, \ldots, n$, hence a lattice in $\mathbb{R}^n$. The fundamental domain is a parallelotope whose sides are the $\sigma(x_i)$, and the volume of the fundamental domain is the absolute value of the determinant whose rows (or columns) are the $\sigma(x_i)$. Consequently [see (5.1.1) for notation],

$$v(\sigma(M)) = |\det C| = 2^{-r_2} |\det \sigma_j(x_k)|.$$

We apply this result in an algebraic number theory setting.

5.3.3 Proposition

Let B be the ring of algebraic integers of a number field L, and let I be a nonzero integral ideal of B, so that by (4.2.4) and (5.3.2), $\sigma(I)$ is a lattice in $\mathbb{R}^n$. Then the volume of the fundamental domain of this lattice is

$$v(\sigma(I)) = 2^{-r_2} |d|^{1/2} N(I);$$

in particular, $v(\sigma(B)) = 2^{-r_2} |d|^{1/2}$, where d is the field discriminant.

Proof. The result for $I = B$ follows from (5.3.2) and (2.3.3), taking the x_k as an integral basis for B. To establish the general result, observe that the fundamental domain for $\sigma(I)$ can be assembled by taking the disjoint union of $N(I)$ copies of the fundamental domain of $\sigma(B)$. To convince yourself of this, let e_1 and e_2 be basis vectors in the plane. The lattice H' generated by $2e_1$ and $3e_2$ is a subgroup of the lattice H generated by e_1 and e_2, but the fundamental domain T' of H' is larger than the fundamental domain T of H. In fact, exactly 6 copies of T will fit inside T'. ♣

5.3.4 Minkowski Bound on Element Norms

If I is a nonzero integral ideal of B, then I contains a nonzero element x such that

$$|N_{L/\mathbb{Q}}(x)| \leq (4/\pi)^{r_2} (n!/n^n) |d|^{1/2} N(I).$$

Proof. The set B_t of Section 5.2 is compact, convex and symmetric about the origin. The volume of B_t is $\mu(B_t) = 2^{r_1} (\pi/2)^{r_2} t^n / n!$, with μ indicating Lebesgue measure. We choose t so that $\mu(B_t) = 2^n v(\sigma(I))$, which by (5.3.3) is $2^{n-r_2}|d|^{1/2}N(I)$. Equating the two expressions for $\mu(B_t)$, we get

$$t^n = 2^{n-r_1}\pi^{-r_2}\ n!\ |d|^{1/2}N(I).$$

Apply (5.1.3b) with $H = \sigma(I)$ and $S = B_t$. By our choice of t, the hypothesis of (5.1.3b) is satisfied, and we have $S \cap (H \setminus \{0\}) \neq \emptyset$. Thus there is a nonzero element $x \in I$ such that $\sigma(x) \in B_t$. Now by (2.1.6), the absolute value of the norm of x is the product of the positive numbers $a_i = |\sigma_i(x)|, i = 1, \ldots, n$. To estimate $N(x)$, we invoke the inequality of the arithmetic and geometric means, which states that $(a_1 \cdots a_n)^{1/n} \le (a_1 + \cdots + a_n)/n$. It follows that $a_1 \cdots a_n \le (\sum_{i=1}^n a_i/n)^n$. With our a_i's, we have

$$|N(x)| \le [\ \frac{1}{n}\sum_{i=1}^{r_1} |\sigma_i(x)| + \frac{2}{n}\sum_{i=r_1+1}^{r_1+r_2} |\sigma_i(x)|\ ]^n.$$

Since $\sigma(x) \in B_t$, we have $|N(x)| \le t^n/n^n$. By choice of t,

$$|N(x)| \le (1/n^n)2^{n-r_1}\pi^{-r_2}\ n!\ |d|^{1/2}N(I).$$

But $n - r_1 = 2r_2$, so $2^{n-r_1}\pi^{-r_2} = 2^{2r_2}\pi^{-r_2} = (4/\pi)^{r_2}$, and the result follows. ♣

5.3.5   Minkowski Bound on Ideal Norms

Every ideal class [see (3.4.5)] of L contains an integral ideal I such that

$$N(I) \le (4/\pi)^{r_2}\ (n!/n^n)\ |d|^{1/2}.$$

Proof. Let J' be a fractional ideal in the given class. We can multiply by a principal ideal of B without changing the ideal class, so we can assume with loss of generality that $J = (J')^{-1}$ is an integral ideal. Choose a nonzero element $x \in J$ such that x satisfies the norm inequality of (5.3.4). Our candidate is $I = xJ'$.

First note that I is an integral ideal because $x \in J$ and $JJ' = B$. Now $(x) = IJ$, so by (4.2.6) and (5.3.4),

$$N(I)N(J) = N(x) \le (4/\pi)^{r_2}\ (n!/n^n)\ |d|^{1/2}N(J).$$

Cancel $N(J)$ to get the desired result. ♣

5.3.6   Corollary

The ideal class group of a number field is finite.

Proof. By (4.2.13), there are only finitely many integral ideals with a given norm. By (5.3.5), we can associate with each ideal class an integral ideal whose norm is bounded above by a fixed constant. If the ideal class group were infinite, we would eventually use the same integral ideal in two different ideal classes, which is impossible. ♣

5.3.7 Applications

Suppose that a number field L has a Minkowski bound on ideal norms that is less than 2. Since the only ideal of norm 1 is the trivial ideal $(1) = B$, every ideal class must contain (1). Thus there can be only one ideal class, and the *class number* of L, that is, the order of the ideal class group, is $h_L = 1$. By (3.4.5), B is a PID, equivalently, by (3.2.8), a UFD.

If the Minkowski bound is greater than 2 but less than 3, we must examine ideals whose norm is 2. If I is such an ideal, then by (4.2.9), I divides (2). Thus the prime factorization of (2) will give useful information about the class number.

In the exercises, we will look at several explicit examples.

Problems For Section 5.3

1. Calculate the Minkowski bound on ideal norms for an imaginary quadratic field, in terms of the field discriminant d. Use the result to show that $\mathbb{Q}(\sqrt{m})$ has class number 1 for $m = -1, -2, -3, -7$.

2. Calculate the Minkowski bound on ideal norms or a real quadratic field, in terms of the field discriminant d. Use the result to show that $\mathbb{Q}(\sqrt{m})$ has class number 1 for $m = 2, 3, 5, 13$.

3. Show that in the ring of algebraic integers of $\mathbb{Q}(\sqrt{-5})$, there is only one ideal whose norm is 2. Then use the Minkowski bound to prove that the class number is 2.

4. Repeat Problem 3 for $\mathbb{Q}(\sqrt{6})$.

5. Show that the only prime ideals of norm 2 in the ring of algebraic integers of $\mathbb{Q}(\sqrt{17})$ are principal. Conclude that the class number is 1.

6. Find the class number of $\mathbb{Q}(\sqrt{14})$. (It will be necessary to determine the number of ideals of norm 3 as well as norm 2.)

Problems 7-10 consider bounds on the field discriminant.

7. Let L be a number field of degree n over $\mathbb{Q}$, with field discriminant d. Show that $|d| \geq a_n = (\pi/4)^n \, n^{2n}/(n!)^2$.

8. Show that $a_2 = \pi^2/4$ and $a_{n+1}/a_n \geq 3\pi/4$. From this, derive the lower bound $|d| \geq (\pi/3)(3\pi/4)^{n-1}$ for $n \geq 2$.

9. Show that $n/\log|d|$ is bounded above by a constant that is independent of the particular number field.

10. Show that if $L \neq \mathbb{Q}$, then $|d| > 1$, hence in any nontrivial extension of $\mathbb{Q}$, at least one prime must ramify.

Chapter 6

The Dirichlet Unit Theorem

As usual, we will be working in the ring B of algebraic integers of a number field L. Two factorizations of an element of B are regarded as essentially the same if one is obtained from the other by multiplication by a unit. Our experience with the integers, where the only units are ± 1, and the Gaussian integers, where the only units are ± 1 and $\pm i$, suggests that units are not very complicated, but this is misleading. The Dirichlet unit theorem gives a complete description of the structure of the multiplicative group of units in a number field.

6.1 Preliminary Results

6.1.1 Lemma

Let B^* be the group of units of B. An element $x \in B$ belongs to B^* if and only if $N(x) = \pm 1$.

Proof. If $xx^{-1} = 1$, then $1 = N(1) = N(xx^{-1}) = N(x)N(x^{-1})$, so the integer $N(x)$ must be ± 1. Conversely, if the norm of x is ± 1, then the characteristic equation of x has the form $x^n + a_{n-1}x^{n-1} + \cdots + a_1 x \pm 1 = 0$, with the $a_i \in \mathbb{Z}$ [see (2.1.3) and (2.2.2)]. Thus $x(x^{n-1} + a_{n-1}x^{n-2} + \cdots + a_2 x + a_1) = \mp 1$. ♣

6.1.2 The Logarithmic Embedding

Let $\sigma : L \to \mathbb{R}^{r_1} \times \mathbb{C}^{r_2} = \mathbb{R}^n$ be the canonical embedding defined in (5.3.1). The *logarithmic embedding* is the mapping $\lambda : L^* \to \mathbb{R}^{r_1+r_2}$ given by

$$\lambda(x) = (\log |\sigma_1(x)|, \ldots, \log |\sigma_{r_1+r_2}(x)|).$$

Since the σ_i are monomorphisms, $\lambda(xy) = \lambda(x) + \lambda(y)$, so λ is a homomorphism from the multiplicative group of L^* to the additive group of $\mathbb{R}^{r_1+r_2}$.

6.1.3   Lemma

Let C be a bounded subset of $\mathbb{R}^{r_1+r_2}$, and let $C' = \{x \in B^* : \lambda(x) \in C\}$. Then C' is a finite set.

Proof. Since C is bounded, all the numbers $|\sigma_i(x)|, x \in B^*, i = 1, \ldots, n$, will be confined to some interval $[a^{-1}, a]$ with $a > 1$. Thus the elementary symmetric functions of the $\sigma_i(x)$ will also lie in some interval of this type. But by (2.1.6), the elementary symmetric functions are the coefficients of the characteristic polynomial of x, and by (2.2.2), these coefficients are integers. Thus there are only finitely many possible characteristic polynomials of elements $x \in C'$, hence by (2.1.5), only finitely many possible roots of minimal polynomials of elements $x \in C'$. We conclude that x can belong to C' for only finitely many x. ♣

6.1.4   Corollary

The kernel G of the homomorphism λ restricted to B^* is a finite group.

Proof. Take $C = \{0\}$ in (6.1.3). ♣

The following result gives additional information about G.

6.1.5   Proposition

Let H be a finite subgroup of K^*, where K is an arbitrary field. Then H consists of roots of unity and is cyclic.

Proof. Let z be an element of H whose order n is the exponent of H, that is, the least common multiple of the orders of all the elements of H. Then $y^n = 1$ for every $y \in H$, so H consists of roots of unity. Since the polynomial $X^n - 1$ has at most n distinct roots, we have $|H| \leq n$. But $1, z, \ldots, z^{n-1}$ are distinct elements of H, because z has order n. Thus H is cyclic. ♣

For our group G, even more is true.

6.1.6   Proposition

The group G consists exactly of all the roots of unity in the field L.

Proof. By (6.1.5), every element of G is a root of unity. Conversely, suppose $x^m = 1$. Then x is an algebraic integer (it satisfies $X^m - 1 = 0$) and for every i,

$$|\sigma_i(x)|^m = |\sigma_i(x^m)| = |1| = 1.$$

Thus $|\sigma_i(x)| = 1$ for all i, so $\log|\sigma_i(x)| = 0$ and $x \in G$. ♣

6.1.7   Proposition

B^* is a finitely generated abelian group, isomorphic to $G \times \mathbb{Z}^s$ where $s \leq r_1 + r_2$.

Proof. By (6.1.3), $\lambda(B^*)$ is a discrete subgroup of $\mathbb{R}^{r_1+r_2}$. ["Discrete" means that any bounded subset of $\mathbb{R}^{r_1+r_2}$ contains only finitely many points of $\lambda(B^*)$.] It follows that

$\lambda(B^*)$ is a lattice in $\mathbb{R}^s$, hence a free $\mathbb{Z}$-module of rank s, for some $s \leq r_1 + r_2$. The proof of this is outlined in the exercises. Now by the first isomorphism theorem, $\lambda(B^*) \cong B^*/G$, with $\lambda(x)$ corresponding to the coset xG. If $x_1 G, \dots, x_s G$ form a basis for B^*/G and $x \in B^*$, then xG is a finite product of powers of the $x_i G$, so x is an element of G times a finite product of powers of the x_i. Since the $\lambda(x_i)$ are linearly independent, so are the x_i, provided we translate the notion of linear independence to a multiplicative setting. The result follows. ♣

We can improve the estimate of s.

6.1.8 Proposition

In (6.1.7), we have $s \leq r_1 + r_2 - 1$.

Proof. If $x \in B^*$, then by (6.1.1) and (2.1.6),

$$\pm 1 = N(x) = \prod_{i=1}^{n} \sigma_i(x) = \prod_{i=1}^{r_1} \sigma_i(x) \prod_{j=r_1+1}^{r_1+r_2} \sigma_j(x)\overline{\sigma_j(x)}.$$

Take absolute values and apply the logarithmic embedding to conclude that $\lambda(x) = (y_1, \dots, y_{r_1+r_2})$ lies in the hyperplane W whose equation is

$$\sum_{i=1}^{r_1} y_i + 2 \sum_{j=r_1+1}^{r_1+r_2} y_j = 0.$$

The hyperplane has dimension $r_1 + r_2 - 1$, so as in the proof of (6.1.7), $\lambda(B^*)$ is a free $\mathbb{Z}$-module of rank $s \leq r_1 + r_2 - 1$. ♣

In the next section, we will prove the Dirichlet unit theorem, which says that s actually equals $r_1 + r_2 - 1$.

Problems For Section 6.1

We will show that if H is a discrete subgroup of $\mathbb{R}^n$, in other words, for every bounded set $C \subseteq \mathbb{R}^n$, $H \cap C$ is finite, then H is a lattice in $\mathbb{R}^r$ for some $r \leq n$. Choose $e_1, \dots, e_r \in H$ such that the e_i are linearly independent over $\mathbb{R}$ and r is as large as possible. Let $\overline{T}$ be the closure of the fundamental domain determined by the e_i, that is, the set of all $x = \sum_{i=1}^{r} a_i e_i$, with $0 \leq a_i \leq 1$. Since H is discrete, $H \cap \overline{T}$ is a finite set.

Now let x be any element of H. By choice of r we have $x = \sum_{i=1}^{r} b_i e_i$ with $b_i \in \mathbb{R}$.

1. If j is any integer, set $x_j = jx - \sum_{i=1}^{r} \lfloor jb_i \rfloor e_i$, where $\lfloor y \rfloor$ is the maximum of all integers $z \leq y$. Show that $x_j \in H \cap T$.
2. By examining the above formula for x_j with $j = 1$, show that H is a finitely generated $\mathbb{Z}$-module.
3. Show that the b_i are rational numbers.
4. Show that for some nonzero integer d, dH is a free $\mathbb{Z}$-module of rank at most r.
5. Show that H is a lattice in $\mathbb{R}^r$.

6.2   Statement and Proof of Dirichlet's Unit Theorem

6.2.1   Theorem

The group B^* of units of a number field L is isomorphic to $G \times \mathbb{Z}^s$, where G is a finite cyclic group consisting of all the roots of unity in L, and $s = r_1 + r_2 - 1$.

Proof. In view of (6.1.4)-(6.1.8), it suffices to prove that $s \geq r_1 + r_2 - 1$. Equivalently, by the proof of (6.1.7), the real vector space $V = \lambda(B^*)$ contains $r_1 + r_2 - 1$ linearly independent vectors. Now by the proof of (6.1.8), V is a subspace of the $(r_1 + r_2 - 1)$-dimensional hyperplane W, so we must prove that $V = W$. To put it another way, every linear form f that vanishes on V must vanish on W. This is equivalent to saying that if f does not vanish on W, then it cannot vanish on V, that is, for some unit $u \in B^*$ we have $f(\lambda(u)) \neq 0$.

Step 1. We apply Minkowski's convex body theorem (5.1.3b) to the set

$$S = \{(y_1, \dots, y_{r_1}, z_1, \dots, z_{r_2}) \in \mathbb{R}^{r_1} \times \mathbb{C}^{r_2} : |y_i| \leq a_i, |z_j| \leq a_{r_1+j}\}$$

where i ranges from 1 to r_1 and j from 1 to r_2. We specify the a_i as follows. Fix the positive real number $b \geq 2^{n-r_1}(1/2\pi)^{r_2}|d|^{1/2}$. Given arbitrary positive real numbers $a_1, \dots, a_r$, where $r = r_1 + r_2 - 1$, we choose the positive real number a_{r+1} such that

$$\prod_{i=1}^{r_1} a_i \prod_{j=r_1+1}^{r_1+r_2} a_j^2 = b.$$

The set S is compact, convex, and symmetric about the origin, and its volume is

$$\prod_{i=1}^{r_1} 2a_i \prod_{j=r_1+1}^{r_1+r_2} \pi a_j^2 = 2^{r_1} \pi^{r_2} b \geq 2^{n-r_2}|d|^{1/2}.$$

We apply (5.1.3b) with S as above and $H = \sigma(B)$ [see (5.3.3)], to get $S \cap (H \setminus \{0\}) \neq \emptyset$. Thus there is a nonzero algebraic integer $x = x_a$, $a = (a_1, \dots, a_r)$, such that $\sigma(x_a) \in S$, and consequently,

$$|\sigma_i(x_a)| \leq a_i, \ i = 1, \dots, n,$$

where we set $a_{j+r_2} = a_j$, $j = r_1 + 1, \dots, r_1 + r_2$.

Step 2. We will show that the norms of the x_a are bounded by b in absolute value, and

$$0 \leq \log a_i - \log|\sigma_i(x_a)| \leq \log b.$$

Using step 1, along with (2.1.6) and the fact that the norm of an algebraic integer is a rational integer [see (2.2.2)], we find

$$1 \leq |N(x_a)| = \prod_{i=1}^{n} |\sigma_i(x_a)| \leq \prod_{i=1}^{r_1} a_i \prod_{j=r_1+1}^{r_1+r_2} a_j^2 = b.$$

But for any i,

$$|\sigma_i(x_a)| = |N(x_a)| \prod_{j\neq i} |\sigma_j(x_a)|^{-1} \geq \prod_{j\neq i} a_j^{-1} = a_i b^{-1}.$$

Thus $a_i b^{-1} \leq |\sigma_i(x_a)| \leq a_i$ for all i, so $1 \leq a_i/|\sigma_i(x_a)| \leq b$. Take logarithms to obtain the desired chain of inequalities.

Step 3. Completion of the proof. In the equation of the hyperplane W, $y_1, \dots, y_r$ can be specified arbitrarily and we can solve for y_{r+1}. Thus if f is a nonzero linear form on W, then f can be expressed as $f(y_1, \dots, y_{r+1}) = c_1 y_1 + \cdots + c_r y_r$ with not all c_i's zero. By definition of the logarithmic embedding [see (6.1.2)], $f(\lambda(x_a)) = \sum_{i=1}^r c_i \log |\sigma_i(x_a)|$, so if we multiply the inequality of Step 2 by c_i and sum over i, we get

$$|\sum_{i=1}^r c_i \log a_i - f(\lambda(x_a))| = |\sum_{i=1}^r c_i(\log a_i - \log |\sigma_i(x_a)|)| \leq \sum_{i=1}^r |c_i| \log b.$$

Choose a positive real number t greater than the right side of this equation, and for every positive integer h, choose positive real numbers $a_{ih}, i = 1, \dots, r$, such that $\sum_{i=1}^r c_i \log a_{ih}$ coincides with $2th$. (This is possible because not all c_i's are zero.) Let $a(h) = (a_{1h}, \dots, a_{rh})$, and let x_h be the corresponding algebraic integer $x_{a(h)}$. Then by the displayed equation above and the choice of t to exceed the right side, we have $|f(\lambda(x_h)) - 2th| < t$, so

$$(2h-1)t < f(\lambda(x_h)) < (2h+1)t.$$

Since the open intervals $((2h-1)t, (2h+1)t)$ are (pairwise) disjoint, it follows that the $f(\lambda(x_h)), h = 1, 2, \dots$, are all distinct. But by Step 2, the norms of the x_h are all bounded in absolute value by the same positive constant, and by (4.2.13), only finitely many ideals can have a given norm. By (4.2.6), there are only finitely many distinct ideals of the form Bx_h, so there are distinct h and k such that $Bx_h = Bx_k$. But then x_h and x_k are associates, hence for some unit u we have $x_h = ux_k$, hence $\lambda(x_h) = \lambda(u) + \lambda(x_k)$. By linearity of f and the fact that $f(\lambda(x_h)) \neq f(\lambda(x_k))$, we have $f(\lambda(u)) \neq 0$. ♣

6.2.2 Remarks

The unit theorem implies that there are $r = r_1 + r_2 - 1$ units $u_1, \dots, u_r$ in B such that every unit of B can be expressed uniquely as

$$u = z\, u_1^{n_1} \cdots u_r^{n_r}$$

where the u_i are algebraic integers and z is a root of unity in L. We call $\{u_1, \dots, u_r\}$ a *fundamental system of units* for the number field L.

As an example, consider the cyclotomic extension $L = \mathbb{Q}(z)$, where z is a primitive p^{th} root of unity, p an odd prime. The degree of the extension is $\varphi(p) = p - 1$, and an embedding σ_j maps z to $z^j, j = 1, \dots, p - 1$. Since these z^j's are never real, we have $r_1 = 0$ and $2r_2 = p - 1$. Therefore $r = r_1 + r_2 - 1 = (p - 3)/2$.

6.3   Units in Quadratic Fields

6.3.1   Imaginary Quadratic Fields

First, we look at number fields $L = \mathbb{Q}(\sqrt{m})$, where m is a square-free negative integer. There are no real embeddings, so $r_1 = 0$ and $2r_2 = n = 2$, hence $r_2 = 1$. But then $r_1 + r_2 - 1 = 0$, so the only units in B are the roots of unity in L. We will use (6.1.1) to determine the units.

Case 1. Assume $m \not\equiv 1 \mod 4$. By (2.3.11), an algebraic integer has the form $x = a + b\sqrt{m}$ for integers a and b. By (6.1.1) and (2.1.10), x is a unit iff $N(x) = a^2 - mb^2 = \pm 1$. Thus if $m \leq -2$, then $b = 0$ and $a = \pm 1$. If $m = -1$, we have the additional possibility $a = 0, b = \pm 1$.

Case 2. Assume $m \equiv 1 \mod 4$. By (2.3.11), $x = a + b(1 + \sqrt{m})/2$, and by (2.1.10), $N(x) = (a + b/2)^2 - mb^2/4 = [(2a + b)^2 - mb^2]/4$. Thus x is a unit if and only if $(2a + b)^2 - mb^2 = 4$. We must examine $m = -3, -7, -11, -15, \ldots$. If $m \leq -7$, then $b = 0, a = \pm 1$. If $m = -3$, we have the additional possibilities $b = \pm 1, (2a \pm b)^2 = 1$, that is, $a = 0, b = \pm 1$; $a = 1, b = -1$; $a = -1, b = 1$.

To summarize, if B is the ring of algebraic integers of an imaginary quadratic field, then the group G of units of B is $\{1, -1\}$, except in the following two cases:
1. If $L = \mathbb{Q}(i)$, then $G = \{1, i, -1, -i\}$, the group of 4th roots of unity in L.
2. If $L = \mathbb{Q}(\sqrt{-3})$, then $G = \{[(1 + \sqrt{-3})/2]^j, \ j = 0, 1, 2, 3, 4, 5\}$, the group of 6th roots of unity in L. We may list the elements $x = a + b/2 + b\sqrt{-3}/2 \in G$ as follows:
$j = 0 \Rightarrow x = 1 \quad (a = 1, b = 0)$
$j = 1 \Rightarrow x = (1 + \sqrt{-3})/2 \quad (a = 0, b = 1)$
$j = 2 \Rightarrow x = (-1 + \sqrt{-3})/2 \quad (a = -1, b = 1)$
$j = 3 \Rightarrow x = -1 \quad (a = -1, b = 0)$
$j = 4 \Rightarrow x = -(1 + \sqrt{-3})/2 \quad (a = 0, b = -1)$
$j = 5 \Rightarrow x = (1 - \sqrt{-3})/2 \quad (a = 1, b = -1)$.

6.3.2   Remarks

Note that G, a finite cyclic group, has a generator, necessarily a primitive root of unity. Thus G will consist of all t^{th} roots of unity for some t, and the field L will contain only finitely many roots of unity. This is a general observation, not restricted to the quadratic case.

6.3.3   Real Quadratic Fields

Now we examine $L = \mathbb{Q}(\sqrt{m})$, where m is a square-free positive integer. Since the $\mathbb{Q}$-automorphisms of L are the identity and $a + b\sqrt{m} \rightarrow a - b\sqrt{m}$, there are two real embeddings and no complex embeddings. Thus $r_1 = 2, r_2 = 0$, and $r_1 + r_2 - 1 = 1$. The only roots of unity in $\mathbb{R}$ are ± 1, so by (6.2.1) or (6.2.2), the group of units in the ring of algebraic integers is isomorphic to $\{-1, 1\} \times \mathbb{Z}$. If u is a unit and $0 < u < 1$, then $1/u$ is a unit and $1/u > 1$. Thus the units greater than 1 are $h^n, n = 1, 2, \ldots$, where h, the unique generator greater than 1, is called the *fundamental unit* of L.

Case 1. Assume $m \not\equiv 1 \mod 4$. The algebraic integers are of the form $x = a + b\sqrt{m}$ with $a, b \in \mathbb{Z}$. Thus we are looking for solutions for $N(x) = a^2 - mb^2 = \pm 1$. Note that if $x = a + b\sqrt{m}$ is a solution, then the four numbers $\pm a \pm b\sqrt{m}$ are $x, -x, x^{-1}, -x^{-1}$ in some order. Since a number and its inverse cannot both be greater than 1, and similarly for a number and its negative, it follows that exactly one of the four numbers is greater than 1, namely the number with a and b positive. The fundamental unit, which is the smallest unit greater than 1, can be found as follows. Compute mb^2 for $b = 1, 2, 3, \ldots$, and stop at the first number mb_1^2 that differs from a square a_1^2 by ± 1. Then $a_1 + b_1\sqrt{m}$ is the fundamental unit.

There is a more efficient computational technique using the continued fraction expansion of $\sqrt{m}$. Details are given in many texts on elementary number theory.

Case 2. Assume $m \equiv 1 \mod 4$. It follows from (2.2.6) that the algebraic integers are of the form $x = \frac{1}{2}(a + b\sqrt{m})$, where a and b are integers of the same parity, both even or both odd. Since the norm of x is $\frac{1}{4}(a^2 - mb^2)$, x is a unit iff $a^2 - mb^2 = \pm 4$. Moreover, if a and b are integers satisfying $a^2 - mb^2 = \pm 4$, then a and b must have the same parity, hence $\frac{1}{2}(a + b\sqrt{m})$ is an algebraic integer and therefore a unit of B. To calculate the fundamental unit, compute $mb^2, b = 1, 2, 3, \ldots$, and stop at the first number mb_1^2 that differs from a square a_1^2 by ± 4. The fundamental unit is $\frac{1}{2}(a_1 + b_1\sqrt{m})$.

Problems For Section 6.3

1. Calculate the fundamental unit of $\mathbb{Q}(\sqrt{m})$ for $m = 2, 3, 5, 6, 7, 10, 11, 13, 14, 15, 17$.

In Problems 2-5, we assume $m \equiv 1 \mod 4$. Suppose that we look for solutions to $a^2 - mb^2 = \pm 1$ (rather than $a^2 - mb^2 = \pm 4$). We get units belonging to a subring $B_0 = \mathbb{Z}[\sqrt{m}]$ of the ring B of algebraic integers, and the positive units of B_0 form a subgroup H of the positive units of B. Let $u = \frac{1}{2}(a + b\sqrt{m})$ be the fundamental unit of the number field L.

2. If a and b are both even, for example when $m = 17$, show that H consists of the powers of u, in other words, $B_0^* = B^*$.

3. If a and b are both odd, show that $u^3 \in B_0$.

4. Continuing Problem 3, show that $u^2 \notin B_0$, so H consists of the powers of u^3.

5. Verify the conclusions of Problems 3 and 4 when $m = 5$ and $m = 13$.

Chapter 7

Cyclotomic Extensions

A cyclotomic extension $\mathbb{Q}(\zeta_n)$ of the rationals is formed by adjoining a primitive n^{th} root of unity ζ_n. In this chapter, we will find an integral basis and calculate the field discriminant.

7.1 Some Preliminary Calculations

7.1.1 The Cyclotomic Polynomial

Recall that the cyclotomic polynomial $\Phi_n(X)$ is defined as the product of the terms $X - \zeta$, where ζ ranges over all primitive n^{th} roots of unity in $\mathbb{C}$. Now an nth root of unity is a primitive d^{th} root of unity for some divisor d of n, so $X^n - 1$ is the product of all cyclotomic polynomials $\Phi_d(X)$ with d a divisor of n. In particular, let $n = p^r$ be a prime power. Since a divisor of p^r is either p^r or a divisor of p^{r-1}, we have

$$\Phi_{p^r}(X) = \frac{X^{p^r} - 1}{X^{p^{r-1}} - 1} = \frac{t^p - 1}{t - 1} = 1 + t + \cdots + t^{p-1}$$

where $t = X^{p^{r-1}}$. If $X = 1$ then $t = 1$, and it follows that $\Phi_{p^r}(1) = p$.

Until otherwise specified, we assume that n is a prime power p^r.

7.1.2 Lemma

Let ζ and ζ' be primitive $(p^r)^{\text{th}}$ roots of unity. Then $u = (1 - \zeta')/(1 - \zeta)$ is a unit in $\mathbb{Z}[\zeta]$, hence in the ring of algebraic integers.

Proof. Since ζ is primitive, $\zeta' = \zeta^s$ for some s (not a multiple of p). It follows that $u = (1 - \zeta^s)/(1 - \zeta) = 1 + \zeta + \cdots + \zeta^{s-1} \in \mathbb{Z}[\zeta]$. By symmetry, $(1 - \zeta)/(1 - \zeta') \in \mathbb{Z}[\zeta'] = \mathbb{Z}[\zeta]$, and the result follows. ♣

7.1.3 Lemma

Let $\pi = 1 - \zeta$ and $e = \varphi(p^r) = p^{r-1}(p - 1)$, where φ is the Euler phi function. Then the principal ideals (p) and $(\pi)^e$ coincide.

Proof. By (7.1.1) and (7.1.2),

$$p = \Phi_{p^r}(1) = \prod_{\zeta'}(1 - \zeta') = \prod_{\zeta'}(\frac{1 - \zeta'}{1 - \zeta})(1 - \zeta) = v(1 - \zeta)^{\varphi(p^r)}$$

where v is a unit in $\mathbb{Z}[\zeta]$. The result follows. ♣

We can now give a short proof of a basic result, but remember that we are operating under the restriction that $n = p^r$.

7.1.4 Proposition

The degree of the extension $\mathbb{Q}(\zeta)/\mathbb{Q}$ equals the degree of the cyclotomic polynomial, namely $\varphi(p^r)$. Therefore the cyclotomic polynomial is irreducible over $\mathbb{Q}$.

Proof. By (7.1.3), p has at least $e = \varphi(p^r)$ prime factors (not necessarily distinct) in the ring of algebraic integers of $\mathbb{Q}(\zeta)$. By the ram-rel identity (4.1.6), $e \leq [\mathbb{Q}(\zeta) : \mathbb{Q}]$. But $[\mathbb{Q}(\zeta) : \mathbb{Q}]$ cannot exceed the degree of a polynomial having ζ as a root, so $[\mathbb{Q}(\zeta) : \mathbb{Q}] \leq e$. If ζ were a root of an irreducible factor of Φ_{p^r}, then the degree of the cyclotomic extension would be less than $\varphi(p^r)$, contradicting what we have just proved. ♣

7.1.5 Lemma

Let B be the ring of algebraic integers of $\mathbb{Q}(\zeta)$. Then (π) is a prime ideal (equivalently, π is a prime element) of B. The relative degree f of (π) over (p) is 1, hence the injection $\mathbb{Z}/(p) \rightarrow B/(\pi)$ is an isomorphism.

Proof. If (π) were not prime, (p) would have more than $\varphi(p^r)$ prime ideal factors, which is impossible, in view of the ram-rel identity. This identity also gives $f = 1$. ♣

We will need to do several discriminant computations, and to prepare for this, we do some calculations of norms. The symbol N with no subscript will mean the norm in the extension $\mathbb{Q}(\zeta)/\mathbb{Q}$.

7.1.6 Proposition

$N(1 - \zeta) = \pm p$, and more generally, $N(1 - \zeta^{p^s}) = \pm p^{p^s}$, $0 \leq s < r$.

Proof. The minimal polynomial of $1-\zeta$ is $\Phi_{p^r}(1-X)$, which has constant term $\Phi_{p^r}(1-0) = p$ by (7.1.1). This proves the first assertion. If $0 < s < r$, then ζ^{p^s} is a primitive $(p^{r-s})^{\text{th}}$ root of unity, so by the above calculation with r replaced by $r - s$,

$$N_1(1 - \zeta^{p^s}) = \pm p$$

where N_1 is the norm in the extension $\mathbb{Q}(\zeta^{p^s})/\mathbb{Q}$. By transitivity of norms [see (2.1.7)] applied to the chain $\mathbb{Q}(\zeta), \mathbb{Q}(\zeta^{p^s}), \mathbb{Q}$, and the formula in (2.1.3) for the norm of an element of the base field, we get

$$N(1 - \zeta^{p^s}) = N_1((1 - \zeta^{p^s})^b)$$

where $b = [\mathbb{Q}(\zeta) : \mathbb{Q}(\zeta^{p^s})] = \varphi(p^r)/\varphi(p^{r-s}) = p^s$. Thus $N(1 - \zeta^{p^s}) = \pm p^b$, and the result follows. ♣

In (7.1.6), the sign is $(-1)^{\varphi(n)}$; see (2.1.3).

7.1.7 Proposition

Let D be the discriminant of the basis $1, \zeta, \ldots, \zeta^{\varphi(p^r)-1}$. Then $D = \pm p^c$, where $c = p^{r-1}(pr - r - 1)$.

Proof. By (2.3.6), $D = \pm N(\Phi'_{p^r}(\zeta))$. Differentiate the equation

$$(X^{p^{r-1}} - 1)\Phi_{p^r}(X) = X^{p^r} - 1$$

to get

$$(X^{p^{r-1}} - 1)\Phi'_{p^r}(X) + p^{r-1}X^{p^{r-1}-1}\Phi_{p^r}(X) = p^r X^{p^r - 1}.$$

Setting $X = \zeta$ and noting that ζ is a root of Φ_{p^r}, we have

$$(\zeta^{p^{r-1}} - 1)\Phi'_{p^r}(\zeta) + 0 = p^r \zeta^{p^r - 1}.$$

Thus

$$\Phi'_{p^r}(\zeta) = \frac{p^r \zeta^{p^r - 1}}{\zeta^{p^{r-1}} - 1}.$$

The norm of the denominator has been computed in (7.1.6). The norm of ζ is ± 1, as ζ is a root of unity. The norm of p^r is $p^{r\varphi(p^r)} = p^{rp^{r-1}(p-1)}$. By (2.1.3), the norm is multiplicative, so the norm of $\Phi'_{p^r}(\zeta)$ is $\pm p^c$, where

$$c = r(p-1)p^{r-1} - p^{r-1} = p^{r-1}(pr - r - 1). \quad \clubsuit$$

7.1.8 Remarks

In (4.2.5), we related the norm of an ideal I to the field discriminant d and the discriminant $D(z)$ of a basis z for I. It is important to notice that the same argument works if I is replaced by any free $\mathbb{Z}$-module J of rank n. Thus if B is the ring of algebraic integers, then

$$D(z) = |B/J|^2 d.$$

Applying this result with $z = \{1, \zeta, \ldots, \zeta^{\varphi(p^r)-1}\}$ and $J = \mathbb{Z}[\zeta]$, we find that

$$D = |B/\mathbb{Z}[\zeta]|^2 d.$$

Thus if we can show that the powers of ζ form an integral basis, so that $\mathbb{Z}[\zeta] = B$, then in view of (7.1.7), we are able to calculate the field discriminant up to sign. Also, by the exercises in Section 4.2, the only ramified prime is p.

Let $\pi = 1 - \zeta$ as in (7.1.3), and recall the isomorphism $\mathbb{Z}/(p) \to B/(\pi)$ of (7.1.5).

7.1.9   Lemma

For every positive integer m, we have $\mathbb{Z}[\zeta] + p^m B = B$.

Proof. We first prove the identity with p replaced by π. If $b \in B$, then $b + (\pi) = t + (\pi)$ for some integer t, hence $b - t \in (\pi)$. Thus $\mathbb{Z}[\zeta] + \pi B = B$, and consequently $\pi \mathbb{Z}[\zeta] + \pi^2 B = \pi B$. Now iterate: If $b \in B$, then $b = b_1 + b_2$, $b_1 \in \mathbb{Z}[\zeta], b_2 \in \pi B$. Then $b_2 = b_3 + b_4$, $b_3 \in \pi \mathbb{Z}[\zeta] \subseteq \mathbb{Z}[\zeta], b_4 \in \pi^2 B$. Observe that $b = (b_1 + b_3) + b_4$, so $\mathbb{Z}[\zeta] + \pi^2 B = B$. Continue in this fashion to obtain the desired result. Now by (7.1.3), $\pi^{\varphi(p^r)}$ is p times a unit, so if $m = \varphi(p^r)$, we can replace $\pi^m B$ by pB, so that $\mathbb{Z}[\zeta] + pB = B$. But we can iterate this equation exactly as above, and the result follows. ♣

7.1.10   Theorem

The set $\{1, \zeta, \ldots, \zeta^{\varphi(p^r)-1}\}$ is an integral basis for the ring of algebraic integers of $\mathbb{Q}(\zeta_{p^r})$.

Proof. By (7.1.7) and (7.1.8), $|B/\mathbb{Z}[\zeta]|$ is a power of p, so $p^m(B/\mathbb{Z}[\zeta]) = 0$ for sufficiently large m. Therefore $p^m B \subseteq \mathbb{Z}[\zeta]$, hence by (7.1.9), $\mathbb{Z}[\zeta] = B$. ♣

Problems For Section 7.1

This problem set will indicate how to find the sign of the discriminant of the basis $1, \alpha, \ldots, \alpha^{n-1}$ of $L = \mathbb{Q}(\alpha)$, where the minimal polynomial f of α has degree n.

1. Let $c_1, \ldots, c_{r_1}$ be the real conjugates of α, that is, the real roots of f, and let $c_{r_1+1}, \overline{c_{r_1+1}}, \ldots, c_{r_1+r_2}, \overline{c_{r_1+r_2}}$ be the complex (=non-real) conjugates. Show that the sign of the discriminant is the sign of

$$\prod_{i=1}^{r_2} (c_{r_1+i} - \overline{c_{r_1+i}})^2.$$

2. Show that the sign of the discriminant is $(-1)^{r_2}$, where $2r_2$ is the number of complex embeddings.

3. Apply the results to $\alpha = \zeta$, where ζ is a primitive $(p^r)^{\text{th}}$ root of unity. (Note that a nontrivial cyclotomic extension has no real embeddings.)

7.2   An Integral Basis of a Cyclotomic Field

In the previous section, we found that the powers of ζ form an integral basis when ζ is a power of a prime. We will extend the result to all cyclotomic extensions.

7.2.1   Notation and Remarks

Let K and L be number fields of respective degrees m and n over $\mathbb{Q}$, and let KL be the composite of K and L. Then KL consists of all finite sums $\sum a_i b_i$ with $a_i \in K$ and $b_i \in L$. This is because the composite can be formed by adjoining basis elements of $K/\mathbb{Q}$ and $L/\mathbb{Q}$ one at a time, thus allowing an induction argument. Let R, S, T be the algebraic integers of K, L, KL respectively. Define RS as the set of all finite sums $\sum a_i b_i$ with $a_i \in R, b_i \in S$. Then $RS \subseteq T$, but equality does not hold in general. For example, look at $K = \mathbb{Q}(\sqrt{m_1})$ and $L = \mathbb{Q}(\sqrt{m_2})$, where $m_1 \equiv 3 \mod 4$, $m_2 \equiv 3 \mod 4$, hence $m_1 m_2 \equiv 1 \mod 4$.

7.2.2   Lemma

Assume that $[KL : \mathbb{Q}] = mn$. Let σ be an embedding of K in $\mathbb{C}$ and τ an embedding of L in $\mathbb{C}$. Then there is an embedding of KL in $\mathbb{C}$ that restricts to σ on K and to τ on L.

Proof. The embedding σ has $[KL : K] = n$ distinct extensions to embeddings of KL in $\mathbb{C}$, and if two of them agree on L, then they agree on KL (because they coincide with σ on K). This contradicts the fact that the extensions are distinct. Thus we have n embeddings of KL in $\mathbb{C}$ with distinct restrictions to L. But there are only n embeddings of L in $\mathbb{C}$, so one of them must be τ, and the result follows. ♣

7.2.3   Lemma

Again assume $[KL : \mathbb{Q}] = mn$. Let $a_1, \ldots, a_m$ and $b_1, \ldots, b_n$ be integral bases for R and S respectively. If $\alpha \in T$, then

$$\alpha = \sum_{i=1}^{m} \sum_{j=1}^{n} \frac{c_{ij}}{r} a_i b_j, \ c_{ij} \in \mathbb{Z}, \ r \in \mathbb{Z}$$

with r having no factor (except ± 1) in common with all the c_{ij}.

Proof. The assumption that $[KL : \mathbb{Q}] = mn$ implies that the $a_i b_j$ form a basis for $KL/\mathbb{Q}$. [See the process of constructing KL discussed in (7.2.1).] In fact the $a_i b_j$ form an integral basis for RS. (This is because RS consists of all finite sums $\sum v_i w_i$, $v_i \in R$, $w_i \in S$. Each v_i is a linear combination of the a_k with integer coefficients, and so on.) It follows that α is a linear combination of the $a_i b_j$ with rational coefficients. Form a common denominator and eliminate common factors to obtain the desired result. ♣

7.2.4   Proposition

We are still assuming that $[KL : \mathbb{Q}] = mn$. If d is the greatest common divisor of the discriminant of R and the discriminant of S, then $T \subseteq \frac{1}{d}RS$. Thus if $d = 1$, then $T = RS$.

Proof. It suffices to show that in (7.2.3), r divides d. To see this, write

$$\frac{c_{ij}}{r} = \frac{c_{ij}(d/r)}{d}.$$

In turn, it suffices to show that r divides the discriminant of R. Then by symmetry, r will also divide the discriminant of S, and therefore divide d.

Let σ be an embedding of K in $\mathbb{C}$. By (7.2.2), σ extends to an embedding (also called σ) of KL in $\mathbb{C}$ such that σ is the identity on L. By (7.2.3), if $\alpha \in T$ we have

$$\sigma(\alpha) = \sum_{i,j} \frac{c_{ij}}{r} \sigma(a_i) b_j.$$

If we set

$$x_i = \sum_{j=1}^{n} \frac{c_{ij}}{r} b_j,$$

we have the system of linear equations

$$\sum_{i=1}^{m} \sigma(a_i) x_i = \sigma(\alpha)$$

where there is one equation for each of the m embeddings σ from K to $\mathbb{C}$. Solving for x_i by Cramer's rule, we get $x_i = \gamma_i/\delta$, where δ is the determinant formed from the $\sigma(a_i)$ and γ_i is the determinant obtained by replacing the ith column of δ with the $\sigma(\alpha)$. Note that by (2.3.3), δ^2 is the discriminant of R, call it e. Since all the $\sigma(a_i)$ and $\sigma(\alpha)$ are algebraic integers, so are δ and all the γ_i. Now

$$x_i = \frac{\gamma_i}{\delta} = \frac{\gamma_i \delta}{\delta^2} = \frac{\gamma_i \delta}{e}$$

so $ex_i = \gamma_i \delta$ is an algebraic integer. By definition of x_i,

$$ex_i = \sum_{j=1}^{n} \frac{ec_{ij}}{r} b_j,$$

an algebraic integer in RS. But e is a $\mathbb{Z}$-linear combination of the a_i, and the $a_i b_j$ are an integral basis for RS, so ec_{ij}/r is an integer. Thus r divides every ec_{ij}. By (7.2.3), r has no factor (except the trivial ± 1) in common with every c_{ij}. Consequently, r divides e, the discriminant of R. ♣

We need one more preliminary result.

7.2.5   Lemma

Let ζ be a primitive n^{th} root of unity, and denote the discriminant of $\{1, \zeta, \ldots, \zeta^{\varphi(n)-1}\}$ by disc(ζ). Then disc(ζ) divides $n^{\varphi(n)}$.

Proof. Let f ($= \Phi_n$, the nth cyclotomic polynomial) be the minimal polynomial of ζ over $\mathbb{Q}$. Since ζ is a root of $X^n - 1$, we have $X^n - 1 = f(X)g(X)$ for some $g \in \mathbb{Q}[X]$. But $f \in \mathbb{Z}[X]$ (because ζ is an algebraic integer), and f, hence g, is monic, so $g \in \mathbb{Z}[X]$. Differentiate both sides of the equation to get $nX^{n-1} = f(X)g'(X) + f'(X)g(X)$. Setting $X = \zeta$, which is a root of f, we have $n\zeta^{n-1} = f'(\zeta)g(\zeta)$. But $\zeta^{n-1} = \zeta^n/\zeta = 1/\zeta$, so

$$n = \zeta f'(\zeta)g(\zeta).$$

Now $[\mathbb{Q}(\zeta) : \mathbb{Q}] = \varphi(n)$, so taking the norm of each side yields

$$n^{\varphi(n)} = N(f'(\zeta))N(\zeta g(\zeta)).$$

But by (2.3.6), $N(f'(\zeta)) = \pm \text{disc}(\zeta)$, and $N(\zeta g(\zeta)) \in \mathbb{Z}$ by (2.2.2). The desired result follows. ♣

7.2.6 Theorem

If ζ is a primitive n^{th} root of unity, then the ring of algebraic integers of $\mathbb{Q}(\zeta)$ is $\mathbb{Z}[\zeta]$. in other words, the powers of ζ form an integral basis.

Proof. We have proved this when ζ is a prime power, so let $n = m_1 m_2$ where the m_i are relatively prime and greater than 1. Now

$$\zeta^{m_1} = (e^{i2\pi/n})^{m_1} = e^{i2\pi m_1/n} = e^{i2\pi/m_2} = \zeta_2,$$

a primitive $(m_2)^{th}$ root of unity, and similarly $\zeta^{m_2} = \zeta_1$, a primitive $(m_1)^{th}$ root of unity. Thus $\mathbb{Q}(\zeta_1)$ and $\mathbb{Q}(\zeta_2)$ are contained in $\mathbb{Q}(\zeta)$. On the other hand, since m_1 and m_2 are relatively prime, there are integers r, s such that $rm_2 + sm_1 = 1$. Thus

$$\zeta = \zeta^{rm_2 + sm_1} = \zeta_1^r \zeta_2^s.$$

It follows that $\mathbb{Q}(\zeta) = \mathbb{Q}(\zeta_1)\mathbb{Q}(\zeta_2)$, and we can apply (7.2.4). In that proposition, we take $K = \mathbb{Q}(\zeta_1), L = \mathbb{Q}(\zeta_2), KL = \mathbb{Q}(\zeta), R = \mathbb{Z}[\zeta_1], S = \mathbb{Z}[\zeta_2]$ (induction hypothesis), $T = RS$. The hypothesis on the degree $[KL : \mathbb{Q}]$ is satisfied because $\varphi(n) = \varphi(m_1)\varphi(m_2)$. By (7.2.5), disc$(\zeta_1)$ divides a power of m_1 and disc(ζ_2) divides a power of m_2. Thus the greatest common divisor of disc(R) and disc(S) is 1, and again the hypothesis of (7.2.4) is satisfied. The conclusion is that the ring T of algebraic integers of KL coincides with RS. But the above argument that $\mathbb{Q}(\zeta) = \mathbb{Q}(\zeta_1)\mathbb{Q}(\zeta_2)$ may be repeated verbatim with $\mathbb{Q}$ replaced by $\mathbb{Z}$. We conclude that $\mathbb{Z}[\zeta] = \mathbb{Z}[\zeta_1]\mathbb{Z}[\zeta_2] = RS = T$. ♣

7.2.7 The Discriminant of a General Cyclotomic Extension

The field discriminant of $\mathbb{Q}(\zeta)$, where ζ is a primitive n^{th} root of unity, is given by

$$\frac{(-1)^{\varphi(n)/2} n^{\varphi(n)}}{\prod_{p|n} p^{\varphi(n)/(p-1)}}.$$

A direct verification, with the aid of (7.1.7) and Problem 3 of Section 7.1, shows that the formula is correct when $n = p^r$. The general case is handled by induction, but the computation is very messy.

In the next chapter, we will study factorization of primes in Galois extensions. The results will apply, in particular, to cyclotomic extensions.

Chapter 8

Factoring of Prime Ideals in Galois Extensions

8.1 Decomposition and Inertia Groups

We return to the general AKLB setup: A is a Dedekind domain with fraction field K, L is a finite separable extension of K, and B is the integral closure of A in L. But now we add the condition that the extension L/K is normal, hence Galois. We will see shortly that the Galois assumption imposes a severe constraint on the numbers e_i and f_i in the ram-rel identity (4.1.6). Throughout this chapter, G will denote the Galois group $\mathrm{Gal}(L/K)$.

8.1.1 Proposition

If $\sigma \in G$, then $\sigma(B) = B$. If Q is a prime ideal of B, then so is $\sigma(Q)$. Moreover, if Q lies above the nonzero prime ideal P of A, then so does $\sigma(Q)$. Thus G acts on the set of prime ideals lying above P.

Proof. If $x \in B$, then $\sigma(x) \in B$ (apply σ to an equation of integral dependence). Thus $\sigma(B) \subseteq B$. But $\sigma^{-1}(B)$ is also contained in B, hence $B = \sigma\sigma^{-1}(B) \subseteq \sigma(B)$. If $PB = \prod Q_i^{e_i}$, then apply σ to get $PB = \prod \sigma(Q_i)^{e_i}$. The $\sigma(Q_i)$ must be prime ideals because σ preserves all algebraic relations. Note also that σ is a K-automorphism, hence fixes every element of A (and of P). Therefore $Q \cap A = P \Rightarrow \sigma(Q) \cap A = P$. ♣

We now show that the action of G is transitive.

8.1.2 Theorem

Let Q and Q_1 be prime ideals lying above P. Then for some $\sigma \in G$ we have $\sigma(Q) = Q_1$.

Proof. If the assertion is false, then for each σ, the ideals Q_1 and $\sigma(Q)$ are maximal and distinct, so $Q_1 \not\subseteq \sigma(Q)$. By the prime avoidance lemma (Section 3.1, exercises), there is an element $x \in Q_1$ belonging to none of the $\sigma(Q)$. Computing the norm of x relative to L/K, we have $N(x) = \prod_{\sigma \in G} \sigma(x)$ by (2.1.6). But one of the σ's is the identity, Q_1 is an ideal, and [by (8.1.1)] $\sigma(x) \in B$ for all σ. Consequently, $N(x) \in Q_1$. But $N(x) \in A$ by

(2.2.2), so $N(x) \in Q_1 \cap A = P = Q \cap A$. Thus $N(x)$ belongs to the prime ideal Q, and therefore some $\sigma^{-1}(x)$ belongs to Q as well. This gives $x \in \sigma(Q)$, a contradiction. ♣

8.1.3 Corollary

In the factorization $PB = \prod_{i=1}^{g} P_i^{e_i}$ of the nonzero prime ideal P, the ramification indices e_i are the same for all i, as are the relative degrees f_i. Thus the ram-rel identity simplifies to $efg = n$, where $n = [L : K] = |G|$.

Proof. This follows from (8.1.2), along with the observation that an automorphism σ preserves all algebraic relations. ♣

Since we have a group G acting on the prime factors of PB, it is natural to consider the stabilizer subgroup of each prime factor Q.

8.1.4 Definitions and Comments

We say that the prime ideals $\sigma(Q), \sigma \in G$, are the *conjugates* of Q. Thus (8.1.2) says that all prime factors of PB are conjugate. The *decomposition group* of Q is the subgroup D of G consisting of those $\sigma \in G$ such that $\sigma(Q) = Q$. (This does *not* mean that σ fixes every element of Q.) By the orbit-stabilizer theorem, the size of the orbit of Q is the index of the stabilizer subgroup D. Since there is only one orbit, of size g,

$$g = [G : D] = |G|/|D|, \text{ hence } |D| = n/g = efg/g = ef,$$

independent of Q. Note also that distinct conjugates of Q determine distinct cosets of D. For if $\sigma_1 D = \sigma_2 D$, then $\sigma_2^{-1}\sigma_1 \in D$, so $\sigma_1(Q) = \sigma_2(Q)$.

There is a particular subgroup of D that will be of interest. By (8.1.1), $\sigma(B) = B$ for every $\sigma \in G$. If $\sigma \in D$, then $\sigma(Q) = Q$. It follows that σ induces an automorphism $\bar{\sigma}$ of B/Q. (Note that $x \equiv y \mod Q$ iff $\sigma x \equiv \sigma y \mod Q$.) Since σ is a K-automorphism, $\bar{\sigma}$ is an A/P-automorphism. The mapping $\sigma \to \bar{\sigma}$ is a group homomorphism from D to the group of A/P-automorphisms of B/Q.

8.1.5 Definition

The kernel I of the above homomorphism, that is, the set of all $\sigma \in D$ such that $\bar{\sigma}$ is trivial, is called the *inertia group* of Q.

8.1.6 Remarks

The inertia group is a normal subgroup of the decomposition group, as it is the kernel of a homomorphism. It is given explicitly by

$$I = \{\sigma \in D : \sigma(x) + Q = x + Q \; \forall x \in B\} = \{\sigma \in D : \sigma(x) - x \in Q \; \forall x \in B\}.$$

We now introduce an intermediate field and ring into the basic $AKLB$ setup, as follows.

Take K_D to be the fixed field of D, and let $A_D = B \cap K_D$ be the integral closure of A in K_D. Let P_D be the prime ideal $Q \cap A_D$. Note that Q is the only prime factor of $P_D B$. This is because all primes in the factorization are conjugate, and $\sigma(Q) = Q$ for all $\sigma \in D$, by definition of D.

8.1.7 Lemma

Let $P_D B = Q^{e'}$ and $f' = [B/Q : A_D/P_D]$. Then $e' = e$ and $f' = f$. Moreover, $A/P \cong A_D/P_D$.

Proof. First, observe that by the ram-rel identity [see (8.1.3)], $e'f' = [L : K_D]$, which is $|D|$ by the fundamental theorem of Galois theory. But $|D| = ef$ by (8.1.4), so $e'f' = ef$. Now as in (4.1.3)-(4.1.5), $A/P \subseteq A_D/P_D \subseteq B/Q$, so $f' \leq f$. Also, $PA_D \subseteq P_D$, so P_D divides PA_D, hence $P_D B$ divides $PA_D B = PB$. Consequently, $e' \leq e$, and this forces $e' = e$ and $f' = f$. Thus the dimension of B/Q over A_D/P_D is the same as the dimension of B/Q over A/P. Since A/P can be regarded as a subfield of A_D/P_D, the proof is complete. ♣

8.1.8 Theorem

Assume $(B/Q)/(A/P)$ separable. The homomorphism $\sigma \to \overline{\sigma}$ of D to $\mathrm{Gal}[(B/Q)/(A/P)]$ introduced in (8.1.4) is surjective with kernel I. Therefore $\mathrm{Gal}[(B/Q)/(A/P)] \cong D/I$.

Proof. Let $\overline{x}$ be a primitive element of B/Q over A/P. Let $x \in B$ be a representative of $\overline{x}$. Let $h(X) = X^r + a_{r-1} X^{r-1} + \cdots + a_0$ be the minimal polynomial of x over K_D; the coefficients a_i belong to A_D by (2.2.2). The roots of h are all of the form $\sigma(x), \sigma \in D$. (We are working in the extension L/K_D, with Galois group D.) By (8.1.7), if we reduce the coefficients of h mod P_D, the resulting polynomial $\overline{h}(X)$ has coefficients in A/P. The roots of $\overline{h}$ are of the form $\overline{\sigma}(\overline{x}), \sigma \in D$ (because $\overline{x}$ is a primitive element). Since $\sigma \in D$ means that $\sigma(Q) = Q$, all conjugates of $\overline{x}$ over A/P lie in B/Q. By the basic theory of splitting fields, B/Q is a Galois extension of A/P.

To summarize, since every conjugate of $\overline{x}$ over A/P is of the form $\overline{\sigma}(\overline{x})$, every A/P-automorphism of B/Q (necessarily determined by its action on $\overline{x}$), is of the form $\overline{\sigma}$ where $\sigma \in D$. Since $\overline{\sigma}$ is trivial iff $\sigma \in I$, it follows that the map $\sigma \to \overline{\sigma}$ is surjective and has kernel I. ♣

8.1.9   Corollary

The order of I is e. Thus the prime ideal P does not ramify if and only if the inertia group of every prime ideal Q lying over P is trivial.

Proof. By definition of relative degree, the order of $\text{Gal}[(B/Q)/(A/P)]$ is f. By (8.1.4), the order of D is ef. Thus by (8.1.8), the order of I must be e. ♣

Problems For Section 8.1

1. Let $D(Q)$ be the decomposition group of the prime ideal Q. It follows from the definition of stabilizer subgroup that $D(\sigma(Q)) = \sigma D(Q)\sigma^{-1}$ for every $\sigma \in G$. Show that the inertia subgroup also behaves in this manner, that is, $I(\sigma(Q)) = \sigma I(Q)\sigma^{-1}$.

2. If L/K is an abelian extension (the Galois group $G = \text{Gal}(L/K)$ is abelian), show that the groups $D(\sigma(Q)), \sigma \in G$, are all equal, as are the $I(\sigma(Q)), \sigma \in G$. Show also that the groups depend only on the prime ideal P of A.

8.2   The Frobenius Automorphism

In the basic $AKLB$ setup, with L/K a Galois extension, we now assume that K and L are number fields.

8.2.1   Definitions and Comments

Let P be a prime ideal of A that does not ramify in B, and let Q be a prime lying over P. By (8.1.9), the inertia group $I(Q)$ is trivial, so by (8.1.8), $\text{Gal}[(B/Q)/(A/P)]$ is isomorphic to the decomposition group $D(Q)$. But B/Q is a finite extension of the finite field A/P [see (4.1.3)], so the Galois group is cyclic. Moreover, there is a canonical generator given by $x+Q \to x^q+Q, x \in B$, where $q = |A/P|$. Thus we have identified a distinguished element $\sigma \in D(Q)$, called the *Frobenius automorphism*, or simply the *Frobenius*, of Q, relative to the extension L/K. The Frobenius automorphism is determined by the requirement that for every $x \in B$,

$$\sigma(x) \equiv x^q \mod Q.$$

We use the notation $\left[\frac{L/K}{Q}\right]$ for the Frobenius automorphism. The behavior of the Frobenius under conjugation is similar to the behavior of the decomposition group as a whole (see the exercises in Section 8.1).

8.2.2   Proposition

If $\tau \in G$, then $\left[\frac{L/K}{\tau(Q)}\right] = \tau \left[\frac{L/K}{Q}\right] \tau^{-1}$.

Proof. If $x \in B$, then $\left[\frac{L/K}{Q}\right] \tau^{-1}x \equiv (\tau^{-1}x)^q = \tau^{-1}x^q \mod Q$. Apply τ to both sides to conclude that $\tau \left[\frac{L/K}{Q}\right] \tau^{-1}$ satisfies the defining equation for $\left[\frac{L/K}{\tau(Q)}\right]$. Since the Frobenius is determined by its defining equation, the result follows. ♣

8.2.3 Corollary

If L/K is abelian, then $\left[\frac{L/K}{Q}\right]$ depends only on P, and we write the Frobenius automorphism as $\left(\frac{L/K}{P}\right)$, and sometimes call it the *Artin symbol*.

Proof. By (8.2.2), the Frobenius is the same for all conjugate ideals $\tau(Q), \tau \in G$, hence by (8.1.2), for all prime ideals lying over P. ♣

8.2.4 Intermediate Fields

We now introduce an intermediate field between K and L, call it F. We can then lift P to the ring of algebraic integers in F, namely $B \cap F$. A prime ideal lying over P has the form $Q \cap F$, where Q is a prime ideal of PB. We will compare decomposition groups with respect to the fields L and F, with the aid of the identity

$$[B/Q : A/P] = [B/Q : (B \cap F)/(Q \cap F)][(B \cap F)/(Q \cap F) : A/P].$$

The term on the left is the order of the decomposition group of Q over P, denoted by $D(Q, P)$. (We are assuming that P does not ramify, so $e = 1$.) The first term on the right is the order of the decomposition group of Q over $Q \cap F$. The second term on the right is the relative degree of $Q \cap F$ over P, call it f. Thus

$$|D(Q, Q \cap F)| = |D(Q, P)|/f$$

Since $D = D(Q, P)$ is cyclic and is generated by the Frobenius automorphism σ, the unique subgroup of D with order $|D|/f$ is generated by σ^f. Note that $D(Q, Q \cap F)$ is a subgroup of $D(Q, P)$, because $\mathrm{Gal}(L/F)$ is a subgroup of $\mathrm{Gal}(L/K)$. It is natural to expect that the Frobenius automorphism of Q, relative to the extension L/F, is σ^f.

8.2.5 Proposition

$$\left[\frac{L/F}{Q}\right] = \left[\frac{L/K}{Q}\right]^f.$$

Proof. Let $\sigma = \left[\frac{L/K}{Q}\right]$. Then $\sigma \in D$, so $\sigma(Q) = Q$; also $\sigma(x) \equiv x^q \mod Q, x \in B$, where $q = |A/P|$. Thus $\sigma^f(Q) = Q$ and $\sigma^f(x) \equiv x^{q^f}$. Since q^f is the cardinality of the field $(B \cap F)/(Q \cap F)$, the result follows. ♣

8.2.6 Proposition

If the extension F/K is Galois, then the restriction of $\sigma = \left[\frac{L/K}{Q}\right]$ to F is $\left[\frac{F/K}{Q \cap F}\right]$.

Proof. Let σ_1 be the restriction of σ to F. Since $\sigma(Q) = Q$, it follows that $\sigma_1(Q \cap F) = Q \cap F$. (Note that F/K is normal, so σ_1 is an automorphism of F.) Thus σ_1 belongs to $D(Q \cap F, P)$. Since $\sigma(x) \equiv x^q \mod Q$, we have $\sigma_1(x) \equiv x^q \mod (Q \cap F)$, where $q = |A/P|$. Consequently, $\sigma_1 = \left[\frac{F/K}{Q \cap F}\right]$. ♣

8.2.7   Definitions and Comments

We may view the lifting from the base field K to the extension field L as occurring in three distinct steps. Let $\mathcal{F}_D$ be the *decomposition field* of the extension, that is, the fixed field of the decomposition group D, and let $\mathcal{F}_I$ be the *inertia field*, the fixed field of the inertia group I. We have the following diagram:

$$
\begin{array}{c}
L \\
\Big| \; e=|I| \\
\mathcal{F}_I \\
\Big| \; f=|D|/e \\
\mathcal{F}_D \\
\Big| \; g=n/ef \\
K
\end{array}
$$

All ramification takes place at the top (call it level 3), and all splitting at the bottom (level 1). There is inertia in the middle (level 2). Alternatively, the results can be expressed in tabular form:

	e	f	g
Level 1	1	1	g
2	1	f	1
3	e	1	1

As we move up the diagram, we multiply the ramification indices and relative degrees. This is often expressed by saying that e and f are *multiplicative in towers*. The basic point is that if $Q = Q_1^{e_1} \cdots$ and $Q_1 = Q_2^{e_2} \cdots$, then $Q = Q_2^{e_1 e_2} \cdots$. The multiplicativity of f follows because f is a vector space dimension.

8.3   Applications

8.3.1   Cyclotomic Fields

Let ζ be a primitive m^{th} root of unity, and let $L = \mathbb{Q}(\zeta)$ be the corresponding cyclotomic field. (We are in the $AKLB$ setup with $A = \mathbb{Z}$ and $K = \mathbb{Q}$.) Assume that p is a rational prime that does not divide m. Then by (7.2.5) and the exercises for Section 4.2, p is unramified. Thus (p) factors in B as $Q_1 \cdots Q_g$, where the Q_i are distinct prime ideals. Moreover, the relative degree f is the same for all Q_i, because the extension $L/\mathbb{Q}$ is Galois. In order to say more about f, we find the Frobenius automorphism σ explicitly. The defining equation is $\sigma(x) \equiv x^p \mod Q_i$ for all i, and consequently

$$\sigma(\zeta) = \zeta^p.$$

(The idea is that the roots of unity remain distinct when reduced mod Q_i, because the polynomial $X^n - 1$ is separable over $\mathbb{F}_p$.)

Now the order of σ is the size of the decomposition group D, which is f. Thus f is the smallest positive integer such that $\sigma^f(\zeta) = \zeta$. Since ζ is a primitive m^{th} root of unity, we conclude that

$$f \text{ is the smallest positive integer such that } p^f \equiv 1 \mod m.$$

Once we know f, we can find the number of prime factors $g = n/f$, where $n = \varphi(m)$. (We already know that $e = 1$ because p is unramified.)

When p divides m, the analysis is more complicated, and we will only state the result. Say $m = p^a m_1$, where p does not divide m_1. Then f is the smallest positive integer such that $p^f \equiv 1 \mod m_1$. The factorization is $(p) = (Q_1 \cdots Q_g)^e$, with $e = \varphi(p^a)$. The Q_i are distinct prime ideals, each with relative degree f. The number of distinct prime factors is $g = \varphi(m_1)/f$.

We will now give a proof of Gauss' law of quadratic reciprocity.

8.3.2   Proposition

Let q be an odd prime, and let $L = \mathbb{Q}(\zeta_q)$ be the cyclotomic field generated by a primitive q^{th} root of unity. Then L has a unique quadratic subfield F. Explicitly, if $q \equiv 1 \mod 4$, then the quadratic subfield is $\mathbb{Q}(\sqrt{q})$, and if $q \equiv 3 \mod 4$, it is $\mathbb{Q}(\sqrt{-q})$. More compactly, $F = \mathbb{Q}(\sqrt{q^*})$, where $q^* = (-1)^{q-1)/2}q$.

Proof. The Galois group of the extension is cyclic of even order $q - 1$, hence has a unique subgroup of index 2. Therefore L has a unique quadratic subfield. By (7.1.7) and the exercises to Section 7.1, the field discriminant is $d = (-1)^{(q-1)/2}q^{q-2} \in \mathbb{Q}$. But $\sqrt{d} \notin \mathbb{Q}$, because d has an odd number of factors of q. If $q \equiv 1 \mod 4$, then the sign of d is positive and $\mathbb{Q}(\sqrt{d}) = \mathbb{Q}(\sqrt{q})$. Similarly, if $q \equiv 3 \mod 4$, then the sign of d is negative and $\mathbb{Q}(\sqrt{d}) = \mathbb{Q}(\sqrt{-q})$. [Note that the roots of the cyclotomic polynomial belong to L, hence so does $\sqrt{d}$; see (2.3.5).] ♣

8.3.3   Remarks

Let σ_p be the Frobenius automorphism $\left(\frac{F/\mathbb{Q}}{p}\right)$, where F is the unique quadratic subfield of L, and p is an odd prime unequal to q. By (4.3.2), case (a1), if q^* is a quadratic residue mod p, then p splits, so $g = 2$ and therefore $f = 1$. Thus the decomposition group D is trivial, and since σ_p generates D, σ_p is the identity. If q^* is not a quadratic residue mod p, then by (4.3.2), case (a2), p is inert, so $g = 1$, $f = 2$, and σ_p is nontrivial. Since the Galois group of $F/\mathbb{Q}$ has only two elements, it may be identified with $\{1, -1\}$ under multiplication, and we may write (using the standard Legendre symbol) $\sigma_p = (\frac{q^*}{p})$. On the other hand, σ_p is the restriction of $\sigma = \left(\frac{L/\mathbb{Q}}{p}\right)$ to F, by (8.2.6). Thus σ_p is the identity on F iff σ belongs to H, the unique subgroup of $\text{Gal}(L/\mathbb{Q})$ of index 2. This will happen iff σ is a square. Now the Frobenius may be viewed as a lifting of the map $x \to x^p \mod q$. [As in (8.3.1), $\sigma(\zeta_q) = \zeta_q^p$.] Thus σ will belong to H iff p is a quadratic residue mod q. In other words, $\sigma_p = (\frac{p}{q})$.

8.3.4   Quadratic Reciprocity

If p and q are distinct odd primes, then

$$\left(\frac{p}{q}\right) = (-1)^{(p-1)(q-1)/4}\left(\frac{q}{p}\right).$$

Proof. By (8.3.3),

$$\left(\frac{p}{q}\right) = \left(\frac{q^*}{p}\right) = \left(\frac{(-1)^{(q-1)/2}}{p}\right)\left(\frac{q}{p}\right) = \left(\frac{-1}{p}\right)^{(q-1)/2}\left(\frac{q}{p}\right).$$

But by elementary number theory, or by the discussion in the introduction to Chapter 1,

$$\left(\frac{-1}{p}\right) = (-1)^{(p-1)/2},$$

and the result follows. ♣

8.3.5   Remark

Let $L = \mathbb{Q}(\zeta)$, where ζ is a primitive p^{th} root of unity, p prime. As usual, B is the ring of algebraic integers of L. In this case, we can factor (p) in B explicitly. By (7.1.3) and (7.1.5),

$$(p) = (1 - \zeta)^{p-1}.$$

Thus the ramification index $e = p - 1$ coincides with the degree of the extension. We say that p is *totally ramified*.

Chapter 9

Local Fields

The definition of *global field* varies in the literature, but all definitions include our primary source of examples, number fields. The other fields that are of interest in algebraic number theory are the *local fields*, which are complete with respect to a discrete valuation. This terminology will be explained as we go along.

9.1 Absolute Values and Discrete Valuations

9.1.1 Definitions and Comments

An *absolute value* on a field k is a mapping $x \to |x|$ from k to the real numbers, such that for every $x, y \in k$,

1. $|x| \geq 0$, with equality if and only if $x = 0$;
2. $|xy| = |x| \, |y|$;
3. $|x + y| \leq |x| + |y|$.

The absolute value is *nonarchimedean* if the third condition is replaced by a stronger version:

3'. $|x + y| \leq \max(|x|, |y|)$.

As expected, *archimedean* means not nonarchimedean.

The familiar absolute values on the reals and the complex numbers are archimedean. However, our interest will be in nonarchimedean absolute values. Here is where most of them come from.

A *discrete valuation* on k is a surjective map $v : k \to \mathbb{Z} \cup \{\infty\}$, such that for every $x, y \in k$,

(a) $v(x) = \infty$ if and only if $x = 0$;
(b) $v(xy) = v(x) + v(y)$;
(c) $v(x + y) \geq \min(v(x), v(y))$.

A discrete valuation induces a nonarchimedean absolute value via $|x| = c^{v(x)}$, where c is a constant with $0 < c < 1$.

9.1.2   Example

Let A be a Dedekind domain with fraction field K, and let P be a nonzero prime ideal of A. Then [see the proof of (4.1.6)] the localized ring A_P is a discrete valuation ring (DVR) with unique maximal ideal (equivalently, unique nonzero prime ideal) PA_P. Choose a generator π of this ideal; this is possible because a DVR is, in particular, a PID. Now if $x \in K^*$, the set of nonzero elements of K, then by factoring the principal fractional ideal $(x)A_P$, we find that $x = u\pi^n$, where $n \in \mathbb{Z}$ and u is a unit in A_P. We define $v_P(x) = n$, with $v_P(0) = \infty$. We can check that v_P is a discrete valuation, called the P-adic valuation on K. Surjectivity and conditions (a) and (b) follow directly from the definition. To verify (c), let $x = u\pi^m$, $y = v\pi^n$ with $m \geq n$. Then $x + y = (v^{-1}u\pi^{m-n} + 1)v\pi^n$, and since the term in parentheses belongs to A_P, the exponent in its prime factorization will be nonnegative. Therefore $v_P(x + y) \geq n = \min(v_P(x), v_P(y))$.

Now consider the special case $A = \mathbb{Z}$, $K = \mathbb{Q}$, $P = (p)$. If x is rational and $x = p^r a/b$ where neither a nor b is divisible by p, then we get the p-adic valuation on the rationals, given by $v_p(p^r a/b) = r$.

Here are some of the basic properties of nonarchimedean absolute values. It is often convenient to exclude the *trivial absolute value*, given by $|x| = 1$ for $x \neq 0$, and $|0| = 0$. Note also that for *any* absolute value, $|1| = |-1| = 1$, $|-x| = |x|$, and $|x^{-1}| = 1/|x|$ for $x \neq 0$. (Observe that $1 \times 1 = (-1) \times (-1) = x \times x^{-1} = 1$.)

9.1.3   Proposition

Let $|\ \ |$ be a nonarchimedean absolute value on the field K. Let A be the corresponding *valuation ring*, defined as $\{x \in K : |x| \leq 1\}$, and P the *valuation ideal* $\{x \in K : |x| < 1\}$. Then A is a local ring with unique maximal ideal P and fraction field K. If $u \in K$, then u is a unit of A if and only if $|u| = 1$. If the trivial absolute value is excluded, then A is not a field.

Proof.

1. A is a ring, because it is closed under addition, subtraction and multiplication, and contains the identity.
2. K is the fraction field of A, because if z is a nonzero element of K, then either z or its inverse belongs to A.
3. A is a local ring with unique maximal ideal P. It follows from the definition that P is a proper ideal. If Q is any proper ideal of A, then $Q \subseteq P$, because $A \setminus P \subseteq A \setminus Q$. (If $x \in A \setminus P$, then $|x| = 1$, hence $|x^{-1}| = 1$, so $x^{-1} \in A$. Thus $x \in Q$ implies that $xx^{-1} = 1 \in Q$, a contradiction.)
4. If $u \in K$, then u is a unit of A iff $|u| = 1$. For if u and v belong to A and $uv = 1$, then $|u|\,|v| = 1$. But both $|u|$ and $|v|$ are at most 1, hence they must equal 1. Conversely, if $|u| = 1$, then $|u^{-1}| = 1$. But then both u and its inverse belong to A, so u is a unit of A.
5. If $|\ \ |$ is nontrivial, then A is not a field. For if $x \neq 0$ and $|x| \neq 1$, then either $|x| < 1$ and $|x^{-1}| > 1$, or $|x| > 1$ and $|x^{-1}| < 1$. Either way, we have an element of A whose inverse lies outside of A. ♣

9.1.4 Proposition

If the nonarchimedean and nontrivial absolute value $|\ |$ on K is induced by the discrete valuation v, then the valuation ring A is a DVR.

Proof. In view of (9.1.3), we need only show that A is a PID. Choose an element $\pi \in A$ such that $v(\pi) = 1$. If x is a nonzero element of A and $v(x) = n \in \mathbb{Z}$, then $v(x\pi^{-n}) = 0$, so $x\pi^{-n}$ has absolute value 1 and is therefore a unit u by (9.1.3). Thus $x = u\pi^n$. Now if I is any proper ideal of A, then I will contain an element $u\pi^n$ with $|n|$ as small as possible, say $|n| = n_0$. Either π^{n_0} or π^{-n_0} will be a generator of I (but not both since I is proper). We conclude that every ideal of A is principal. ♣

The proof of (9.1.4) shows that A has exactly one nonzero prime ideal, namely (π).

9.1.5 Proposition

If $|\ |$ is a nonarchimedean absolute value , then $|x| \neq |y|$ implies $|x + y| = \max(|x|, |y|)$. Hence by induction, if $|x_1| > |x_i|$ for all $i = 2, \dots, n$, then $|x_1 + \cdots + x_n| = |x_1|$.

Proof. We may assume without loss of generality that $|x| > |y|$. Then

$$|x| = |x + y - y| \leq \max(|x + y|, |y|) = |x + y|,$$

otherwise $\max(|x+y|, |y|) = |y| < |x|$, a contradiction. Since $|x+y| \leq \max(|x|, |y|) = |x|$, the result follows. ♣

9.1.6 Corollary

With respect to the metric induced by a nonarchimedean absolute value, all triangles are isosceles.

Proof. Let the vertices of the triangle be x, y and z. Then $|x - y| = |(x - z) + (z - y)|$. If $|x - z| = |z - y|$, then two side lengths are equal. If $|x - z| \neq |z - y|$, then by (9.1.5), $|x - y| = \max(|x - z|, |z - y|)$, and again two side lengths are equal. ♣

9.1.7 Proposition

The absolute value $|\ |$ is nonarchimedean if and only if $|n| \leq 1$ for every integer $n = 1 \pm \cdots \pm 1$, equivalently if and only if the set $\{|n| : n \in \mathbb{Z}\}$ is bounded.

Proof. If the absolute value is nonarchimedean, then $|n| \leq 1$ by repeated application of condition 3' of (9.1.1). Conversely, if every integer has absolute value at most 1, then it suffices to show that $|x + 1| \leq \max(|x|, 1)$ for every x. (Apply this result to $x/y, y \neq 0$.) By the binomial theorem,

$$|x + 1|^n = \left| \sum_{r=0}^{n} \binom{n}{r} x^r \right| \leq \sum_{r=0}^{n} \left| \binom{n}{r} \right| |x|^r.$$

By hypothesis, the integer $\binom{n}{r}$ has absolute value at most 1. If $|x| > 1$, then $|x|^r \leq |x|^n$ for all $r = 0, 1, \dots, n$. If $|x| \leq 1$, then $|x|^r \leq 1$. Consequently,

$$|x + 1|^n \leq (n + 1) \max(|x|^n, 1).$$

Take n^{th} roots and let $n \to \infty$ to get $|x+1| \leq \max(|x|, 1)$. Finally, to show that bounded-ness of the set of integers is an equivalent condition, note that if $|n| > 1$, then $|n|^j \to \infty$ as $j \to \infty$ ♣

Problems For Section 9.1

1. Show that every absolute value on a finite field is trivial.

2. Show that a field that has an archimedean absolute value must have characteristic 0.

3. Two nontrivial absolute values $| \ |_1$ and $| \ |_2$ on the same field are said to be *equivalent* if for every x, $|x|_1 < 1$ if and only if $|x|_2 < 1$. [Equally well, $|x|_1 > 1$ if and only if $|x|_2 > 1$; just replace x by $1/x$ if $x \neq 0$.] This says that the absolute values induce the same topology (because they have the same sequences that converge to 0). Show that two nontrivial absolute values are equivalent if and only if for some real number a, we have $|x|_1^a = |x|_2$ for all x.

9.2 Absolute Values on the Rationals

In (9.1.2), we discussed the p-adic absolute value on the rationals (induced by the p-adic valuation, with p prime), and we are familiar with the usual absolute value. In this section, we will prove that up to equivalence (see Problem 3 of Section 9.1), there are no other nontrivial absolute values on $\mathbb{Q}$.

9.2.1 Preliminary Calculations

Fix an absolute value $| \ |$ on $\mathbb{Q}$. If m and n are positive integers greater than 1, expand m to the base n. Then $m = a_0 + a_1 n + \cdots + a_r n^r$, $0 \leq a_i \leq n-1$, $a_r \neq 0$.

(1) $r \leq \log m / \log n$.

This follows because $n^r \leq m$.

(2) For every positive integer l we have $|l| \leq l$, hence in the above base n expansion, $|a_i| \leq a_i < n$.

This can be done by induction: $|1| = 1, |1 + 1| \leq |1| + |1|$, and so on.

There are $1 + r$ terms in the expansion of m, each bounded by $n[\max(1, |n|)]^r$. [We must allow for the possibility that $|n| < 1$, so that $|n|^i$ decreases as i increases. In this case, we will not be able to claim that $|a_0| \leq n(|n|^r)$.] With the aid of (1), we have

(3) $|m| \leq (1 + \log m / \log n) n[\max(1, |n|)]^{\log m / \log n}$.

Replace m by m^t and take the tth root of both sides. The result is

(4) $|m| \leq (1 + t \log m / \log n)^{1/t} n^{1/t} [\max(1, |n|)]^{\log m / \log n}$.

Let $t \to \infty$ to obtain our key formula:

(5) $|m| \leq [\max(1, |n|)]^{\log m / \log n}$.

9.2.2 The Archimedean Case

Suppose that $|n| > 1$ for every $n > 1$. Then by (5), $|m| \leq |n|^{\log m / \log n}$, and therefore $\log |m| \leq (\log m / \log n) \log |n|$. Interchanging m and n gives the reverse inequality, so $\log |m| = (\log m / \log n) \log |n|$. It follows that $\log |n| / \log n$ is a constant a, so $|n| = n^a$. Since $1 < |n| \leq n$ [see (2)], we have $0 < a \leq 1$. Thus our absolute value is equivalent to the usual one.

9.2.3 The Nonarchimedean Case

Suppose that for some $n > 1$ we have $|n| \leq 1$. By (5), $|m| \leq 1$ for all $m > 1$, so $|n| \leq 1$ for all $n \geq 1$, and the absolute value is nonarchimedean by (9.1.7). Excluding the trivial absolute value, we have $|n| < 1$ for some $n > 1$. (If every nonzero integer has absolute value 1, then every nonzero rational number has absolute value 1.) Let $P = \{n \in \mathbb{Z} : |n| < 1\}$. Then P is a prime ideal (p). (Note that if ab has absolute value less than 1, so does either a or b.) Let $c = |p|$, so $0 < c < 1$.

Now let r be the exact power of p dividing n, so that p^r divides n but p^{r+1} does not. Then $n/p^r \notin P$, so $|n|/c^r = 1$, $|n| = c^r$. Note that n/p^{r+1} also fails to belong to P, but this causes no difficulty because n/p^{r+1} is not an integer.

To summarize, our absolute value agrees, up to equivalence, with the p-adic absolute value on the positive integers, hence on all rational numbers. (In going from a discrete valuation to an absolute value, we are free to choose any constant in $(0,1)$. A different constant will yield an equivalent absolute value.)

Problems For Section 9.2

If v_p is the p-adic valuation on $\mathbb{Q}$, let $\| \ \|_p$ be the associated absolute value with the particular choice $c = 1/p$. Thus $\|p^r\|_p = p^{-r}$. Denote the usual absolute value by $\| \ \|_\infty$.

1. Establish the *product formula*: If a is a nonzero rational number, then

$$\prod_p \|a\|_p = 1$$

where p ranges over all primes, including the "infinite prime" $p = \infty$.

9.3 Artin-Whaples Approximation Theorem

The Chinese remainder theorem states that if $I_1, \ldots I_n$ are ideals in a ring R that are relatively prime in pairs, and $a_i \in I_i$, $i = 1, \ldots, n$, then there exists $a \in R$ such that $a \equiv a_i \mod I_i$ for all i. We are going to prove a result about mutually equivalent absolute values that is in a sense analogous. The condition $a \equiv a_i \mod I_i$ will be replaced by the statement that a is close to a_i with respect to the ith absolute value. First, some computations.

9.3.1   Lemma

Let $|\ \ |$ be an arbitrary absolute value. Then

(1) $|a| < 1 \Rightarrow a^n \to 0$;

(2) $|a| < 1 \Rightarrow a^n/(1 + a^n) \to 0$;

(3) $|a| > 1 \Rightarrow a^n/(1 + a^n) \to 1$.

Proof. The first statement follows from $|a^n| = |a|^n$. To prove (2), use the triangle inequality and the observation that $1 + a^n = 1 - (-a^n)$ to get

$$1 - |a|^n \le |1 + a^n| \le 1 + |a|^n,$$

so by (1), $|1 + a^n| \to 1$. Since $|\alpha/\beta| = |\alpha|/|\beta|$, another application of (1) gives the desired result. To prove (3), write

$$1 - \frac{a^n}{1 + a^n} = \frac{1}{1 + a^n} = \frac{a^{-n}}{1 + a^{-n}} \to 0 \text{ by (2).} \quad \clubsuit$$

Here is the key step in the development.

9.3.2   Proposition

Let $|\ \ |_1, \dots, |\ \ |_n$ be nontrivial, mutually inequivalent absolute values on the same field. Then there is an element a such that $|a|_1 > 1$ and $|a|_i < 1$ for $i = 2, \dots, n$.

Proof. First consider the case $n = 2$. Since $|\ \ |_1$ and $|\ \ |_2$ are inequivalent, there are elements b and c such that $|b|_1 < 1$, $|b|_2 \ge 1$, $|c|_1 \ge 1$, $|c|_2 < 1$. If $a = c/b$, then $|a|_1 > 1$ and $|a|_2 < 1$.

Now if the result holds for $n - 1$, we can choose an element b such that $|b|_1 > 1, |b|_2 < 1, \dots, |b|_{n-1} < 1$. By the $n = 2$ case, we can choose c such that $|c|_1 > 1$ and $|c|_n < 1$.

Case 1. Suppose $|b|_n \le 1$. Take $a_r = cb^r$, $r \ge 1$. Then $|a_r|_1 > 1$, $|a_r|_n < 1$, and $|a_r|_i \to 0$ as $r \to \infty$ for $i = 2, \dots, n - 1$. Thus we can take $a = a_r$ for sufficiently large r.

Case 2. Suppose $|b|_n > 1$. Take $a_r = cb^r/(1 + b^r)$. By (3) of (9.3.1), $|a_r|_1 \to |c|_1 > 1$ and $|a_r|_n \to |c|_n < 1$ as $r \to \infty$. If $2 \le i \le n - 1$, then $|b|_i < 1$, so by (2) of (9.3.1), $|a_r|_i \to 0$ as $r \to \infty$. Again we can take $a = a_r$ for sufficiently large r. $\quad \clubsuit$

9.3.3   Approximation Theorem

Let $|\ \ |_1, \dots, |\ \ |_n$ be nontrivial mutually inequivalent absolute values on the field k. Given arbitrary elements $x_1, \dots, x_n \in k$ and any positive real number ϵ, there is an element $x \in k$ such that $|x - x_i|_i < \epsilon$ for all $i = 1, \dots, n$.

Proof. By (9.3.2), $\forall i \ \exists y_i \in k$ such that $|y_i|_i > 1$ and $|y_i|_j < 1$ for $j \ne i$. Take $z_i = y_i^r/(1 + y_i^r)$. Given $\delta > 0$, it follows from (2) and (3) of (9.3.1) that for r sufficiently large,

$$|z_i - 1|_i < \delta \text{ and } |z_j| < \delta, \ j \ne i.$$

Our candidate is

$$x = x_1 z_1 + \cdots x_n z_n.$$

To show that x works, note that $x - x_i = \sum_{j \neq i} x_j z_j + x_i(z_i - 1)$. Thus

$$|x - x_i|_i \leq \delta \sum_{j \neq i} |x_j|_i + \delta |x_i|_i = \delta \sum_{j=1}^{n} |x_j|_i \ .$$

Choose δ so that the right side is less than ϵ, and the result follows. ♣

Problems For Section 9.3

1. Let $|\ |_1, \dots, |\ |_n$ be nontrivial mutually inequivalent absolute values on the field k. Fix r with $0 \leq r \leq n$. Show that there is an element $a \in k$ such that $|a|_1 > 1, \dots, |a|_r > 1$ and $|a|_{r+1}, \dots, |a|_n < 1$.

2. There is a gap in the first paragraph of the proof of (9.3.2), which can be repaired by showing that the implication $|a|_1 < 1 \Rightarrow |a|_2 < 1$ is sufficient for equivalence. Prove this.

9.4 Completions

You have probably seen the construction of the real numbers from the rationals, and the general process of completing a metric space using equivalence classes of Cauchy sequences. If the metric is induced by an absolute value on a field, then we have some additional structure that we can exploit to simplify the development. If we complete the rationals with respect to the p-adic rather than the usual absolute value, we get the p-adic numbers, the most popular example of a local field.

9.4.1 Definitions and Comments

Let K be a field with an absolute value $|\ |$, and let C be the set of Cauchy sequences with elements in K. Then C is a ring under componentwise addition and multiplication. Let N be the set of *null sequences* (sequences converging to 0). Then N is an ideal of C (because every Cauchy sequence is bounded). In fact N is a maximal ideal, because every Cauchy sequence not in N is eventually bounded away from 0, hence is a unit in C. The *completion* of K with respect to the given absolute value is the field $\hat{K} = C/N$. We can embed K in $\hat{K}$ via $c \to \{c, c, \dots\} + N$.

We now extend the absolute value on K to $\hat{K}$. If $(c_n) + N \in \hat{K}$, then $(|c_n|)$ is a Cauchy sequence of real numbers, because by the triangle inequality, $|c_n| - |c_m|$ has (ordinary) absolute value at most $|c_n - c_m| \to 0$ as $n, m \to \infty$. Thus $|c_n|$ converges to a limit, which we take as the absolute value of $(c_n) + N$. Since the original absolute value satisfies the defining conditions in (9.1.1), so does the extension.

To simplify the notation, we will denote the element $(c_n) + N$ of $\hat{K}$ by (c_n). If $c_n = c \in K$ for all n, we will write the element as c.

9.4.2 Theorem

K is dense in $\hat{K}$ and $\hat{K}$ is complete.

Proof. Let $\alpha = (c_n) \in \hat{K}$, with $\alpha_n = c_n$. Then

$$|\alpha - \alpha_n| = \lim_{m \to \infty} |c_m - c_n| \to 0 \text{ as } n \to \infty,$$

proving that K is dense in $\hat{K}$. To prove completeness of $\hat{K}$, let (α_n) be a Cauchy sequence in $\hat{K}$. Since K is dense, for every positive integer n there exists $c_n \in K$ such that $|\alpha_n - c_n| < 1/n$. But then (c_n) is a Cauchy sequence in $\hat{K}$, hence in K, and we are assured that $\alpha = (c_n)$ is a legal element of $\hat{K}$. Moreover, $|\alpha_n - \alpha| \to 0$, proving completeness. ♣

9.4.3   Uniqueness of the Completion

Suppose K is isomorphic to a dense subfield of the complete field L, where the absolute value on L extends that of (the isomorphic copy of) K. If $x \in \hat{K}$, then there is a sequence $x_n \in K$ such that $x_n \to x$. But the sequence (x_n) is also Cauchy in L, hence converges to an element $y \in L$. If we define $f(x) = y$, then f is a well-defined homomorphism of fields, necessarily injective. If $y \in L$, then y is the limit of a Cauchy sequence in K, which converges to some $x \in \hat{K}$. Consequently, $f(x) = y$. Thus f is an isomorphism of $\hat{K}$ and L, and f preserves the absolute value.

9.4.4   Power Series Representation

We define a *local field* K as follows. There is an absolute value on K induced by a discrete valuation v, and with respect to this absolute value, K is complete. For short, we say that K is complete with respect to the discrete valuation v. Let A be the valuation ring (a DVR), and P the valuation ideal; see (9.1.3) and (9.1.4) for terminology. If $\alpha \in K$, then by (9.1.4) we can write $\alpha = u\pi^r$ with $r \in \mathbb{Z}, u$ a unit in A and π an element of A such that $v(\pi) = 1$. Often, π is called a *prime element* or a *uniformizer*. Note that $A = \{\alpha \in K : v(\alpha) \geq 0\}$ and $P = \{\alpha \in K : v(\alpha) \geq 1\} = A\pi$.

Let S be a fixed set of representatives of the cosets of A/P. We will show that each $\alpha \in K$ has a Laurent series expansion

$$\alpha = a_{-m}\pi^{-m} + \cdots + a_{-1}\pi^{-1} + a_0 + a_1\pi + a_2\pi^2 + \cdots, a_i \in S,$$

and if a_r is the first nonzero coefficient (r may be negative), then $v(\alpha) = r$.

The idea is to expand the unit u in a power series involving only nonnegative powers of π. For some $a_0 \in S$ we have $u - a_0 \in P$. But then $v(u - a_0) \geq 1$, hence $v((u - a_0)/\pi) \geq 0$, so $(u - a_0)/\pi \in A$. Then for some $a_1 \in S$ we have $[(u - a_0)/\pi] - a_1 \in P$, in other words,

$$\frac{u - a_0 - a_1\pi}{\pi} \in P.$$

Repeating the above argument, we get

$$\frac{u - a_0 - a_1\pi}{\pi^2} \in A.$$

Continue inductively to obtain the desired series expansion. Note that by definition of S, the coefficients a_i are unique. Thus an expansion of α that begins with a term of degree r in π corresponds to a representation $\alpha = u\pi^r$ and a valuation $v(\alpha) = r$. Also, since $|\pi| < 1$, high positive powers of π are small with respect to the given absolute value. The partial sums s_n of the series form a *coherent sequence*, that is, $s_n \equiv s_{n-1} \mod (\pi)^n$.

9.4.5 Proposition

Let $\sum a_n$ be any series of elements in a local field. Then the series converges if and only if $a_n \to 0$.

Proof. If the series converges, then $a_n \to 0$ by the standard calculus argument, so assume that $a_n \to 0$. Since the absolute value is nonarchimedean, $n \leq m$ implies that

$$|\sum_{i=n}^{m} a_i| \leq \max(a_n, \dots, a_m) \to 0 \text{ as } n \to \infty. \quad \clubsuit$$

9.4.6 Definitions and Comments

The completion of the rationals with respect to the p-adic valuation is called the field of *p-adic numbers*, denoted by $\mathbb{Q}_p$. The valuation ring $A = \{\alpha : v(\alpha) \geq 0\}$ is called the ring of *p-adic integers*, denoted by $\mathbb{Z}_p$. The series representation of a p-adic integer contains only nonnegative powers of $\pi = p$. If in addition, there is no constant term, we get the valuation ideal $P = \{\alpha : v(\alpha) \geq 1\}$. The set S of coset representatives may be chosen to be $\{0, 1, \dots, p-1\}$. (Note that if $a \neq b$ and $a \equiv b \mod p$, then $a - b \in P$, so a and b cannot both belong to S. Also, a rational number can always be replaced by an integer with the same valuation.) Arithmetic is carried out via polynomial multiplication, except that there is a "carry". For example, if $p = 7$, then $3 + 6 = 9 = 2 + p$. For some practice, see the exercises.

We adopt the convention that in going from the p-adic valuation to the associated absolute value $|x| = c^{v(x)}, 0 < c < 1$, we take $c = 1/p$. Thus $|p^r| = p^{-r}$.

Problems For Section 9.4

1. Show that a rational number a/b (in lowest terms) is a p-adic integer if and only if p does not divide b.
2. With $p = 3$, express the product of $(2 + p + p^2)$ and $(2 + p^2)$ as a p-adic integer.
3. Express the p-adic integer -1 as an infinite series.
4. Show that the sequence $a_n = n!$ of p-adic integers converges to 0.
5. Does the sequence $a_n = n$ of p-adic integers converge?
6. Show that the p-adic power series for $\log(1+x)$, namely $\sum_{n=1}^{\infty}(-1)^{n+1}x^n/n$, converges in $\mathbb{Q}_p$ for $|x| < 1$ and diverges elsewhere. This allows a definition of a p-adic logarithm: $\log_p(x) = \log[1 + (x - 1)]$.

In Problems 7-9, we consider the p-adic exponential function.

7. Recall from elementary number theory that the highest power of p dividing $n!$ is $\sum_{i=1}^{\infty}\lfloor n/p^i \rfloor$. (As an example, let $n = 15$ and $p = 2$. Calculate the number of multiples of 2, 4, and 8 in the integers 1-15.) Use this result to show that the p-adic valuation of $n!$ is at most $n/(p-1)$.
8. Show that the p-adic valuation of $(p^m)!$ is $(p^m - 1)/(p - 1)$.
9. Show that the exponential series $\sum_{n=0}^{\infty} x^n/n!$ converges for $|x| < p^{-1/(p-1)}$ and diverges elsewhere.

9.5   Hensel's Lemma

9.5.1   The Setup

Let K be a local field with valuation ring A and valuation ideal P. By (9.1.3) and (9.1.4), A is a local ring, in fact a DVR, with maximal ideal P. The field $k = A/P$ is called the *residue field* of A or of K. If $a \in A$, then the coset $a + P \in k$ will be denoted by $\bar{a}$. If f is a polynomial in $A[X]$, then reduction of the coefficients of f mod P yields a polynomial $\overline{f}$ in $k[X]$. Thus

$$f(X) = \sum_{i=0}^{d} a_i X^i \in A[X], \ \overline{f}(X) = \sum_{i=0}^{d} \overline{a}_i X^i \in k[X].$$

Hensel's lemma is about lifting a factorization of $\overline{f}$ from $k[X]$ to $A[X]$. Here is the precise statement.

9.5.2   Hensel's Lemma

Assume that f is a monic polynomial of degree d in $A[X]$, and that the corresponding polynomial $F = \overline{f}$ factors as the product of relatively prime monic polynomials G and H in $k[X]$. Then there are monic polynomials g and h in $A[X]$ such that $\bar{g} = G$, $\bar{h} = H$ and $f = gh$.

Proof. Let r be the degree of G, so that $\deg H = d - r$. We will inductively construct $g_n, h_n \in A[X], n = 1, 2, \ldots$, such that $\deg g_n = r$, $\deg h_n = d - r$, $\bar{g}_n = G$, $\bar{h}_n = H$, and

$$f(X) - g_n(X)h_n(X) \in P^n[X].$$

Thus the coefficients of $f - g_n h_n$ belong to P^n.

The basis step: Let $n = 1$. Choose monic $g_1, h_1 \in A[X]$ such that $\bar{g}_1 = G$ and $\bar{h}_1 = H$. Then $\deg g_1 = r$ and $\deg h_1 = d - r$. Since $\overline{f} = \bar{g}_1 \bar{h}_1$, we have $f - g_1 h_1 \in P[X]$.

The inductive step: Assume that g_n and h_n have been constructed. Let $f(X) - g_n(X)h_n(X) = \sum_{i=0}^{d} c_i X^i$ with the $c_i \in P^n$. Since $G = \bar{g}_n$ and $H = \bar{h}_n$ are relatively prime, for each $i = 0, \ldots, d$ there are polynomials $\overline{v}_i$ and $\overline{w}_i$ in $k[X]$ such that

$$X^i = \overline{v}_i(X)\bar{g}_n(X) + \overline{w}_i(X)\bar{h}_n(X).$$

Since $\bar{g}_n$ has degree r, the degree of $\overline{v}_i$ is at most $d - r$, and similarly the degree of $\overline{w}_i$ is at most r. Moreover,

$$X^i - v_i(X)g_n(X) - w_i(X)h_n(X) \in P[X]. \tag{1}$$

We define

$$g_{n+1}(X) = g_n(X) + \sum_{i=0}^{d} c_i w_i(X), \ h_{n+1}(X) = h_n(X) + \sum_{i=0}^{d} c_i v_i(X).$$

Since the c_i belong to $P^n \subseteq P$, it follows that $\overline{g}_{n+1} = \overline{g}_n = G$ and $\overline{h}_{n+1} = \overline{h}_n = H$. Since the degree of g_{n+1} is at most r, it must be exactly r, and similarly the degree of h_{n+1} is $d - r$. To check the remaining condition,

$$f - g_{n+1}h_{n+1} = f - (g_n + \sum_i c_i w_i)(h_n + \sum_i c_i v_i)$$

$$= (f - g_n h_n - \sum_i c_i X^i) + \sum_i c_i(X^i - g_n v_i - h_n w_i) - \sum_{i,j} c_i c_j w_i v_j.$$

By the induction hypothesis, the first grouped term on the right is zero, and, with the aid of Equation (1) above, the second grouped term belongs to $P^n P[X] = P^{n+1}[X]$. The final term belongs to $P^{2n}[X] \subseteq P^{n+1}[X]$, completing the induction.

Finishing the proof. By definition of g_{n+1}, we have $g_{n+1} - g_n \in P^n[X]$, so for any fixed i, the sequence of coefficients of X^i in $g_n(X)$ is Cauchy and therefore converges. To simplify the notation we write $g_n(X) \to g(X)$, and similarly $h_n(X) \to h(X)$, with $g(X), h(X) \in A[X]$. By construction, $f - g_n h_n \in P^n[X]$, and we may let $n \to \infty$ to get $f = gh$. Since $\overline{g}_n = G$ and $\overline{h}_n = H$ for all n, we must have $\overline{g} = G$ and $\overline{h} = H$. Since f, G and H are monic, the highest degree terms of g and h are of the form $(1 + a)X^r$ and $(1 + a)^{-1}X^{d-r}$ respectively, with $a \in P$. (Note that $1 + a$ must reduce to 1 mod P.) By replacing g and h by $(1 + a)^{-1}g$ and $(1 + a)h$, respectively, we can make g and h monic without disturbing the other conditions. The proof is complete. ♣

9.5.3 Corollary

With notation as in (9.5.1), let f be a monic polynomial in $A[X]$ such that $\overline{f}$ has a simple root $\eta \in k$. Then f has a simple root $a \in A$ such that $\overline{a} = \eta$.

Proof. We may write $\overline{f}(X) = (X - \eta)H(X)$ where $X - \eta$ and $H(X)$ are relatively prime in $k[X]$. By Hensel's lemma, we may lift the factorization to $f(X) = (X - a)h(X)$ with $h \in A[X], a \in A$ and $\overline{a} = \eta$. If a is a multiple root of f, then η is a multiple root of $\overline{f}$, which is a contradiction. ♣

Problems For Section 9.5

1. Show that for any prime p, there are $p - 1$ distinct $(p - 1)^{\text{th}}$ roots of unity in $\mathbb{Z}_p$.

2. Let p be an odd prime not dividing the integer m. We wish to determine whether m is a square in $\mathbb{Z}_p$. Describe an effective procedure for doing this.

3. In Problem 2, suppose that we not only want to decide if m is a square in $\mathbb{Z}_p$, but to find the series representation of $\sqrt{m}$ explicitly. Indicate how to do this, and illustrate with an example.

Appendices

We collect some results that might be covered in a first course in algebraic number theory.

A. Quadratic Reciprocity Via Gauss Sums

A1. Introduction

In this appendix, p is an odd prime unless otherwise specified. A quadratic equation modulo p looks like $ax^2 + bx + c = 0$ in $\mathbb{F}_p$. Multiplying by $4a$, we have

$$\left(2ax + b\right)^2 \equiv b^2 - 4ac \mod p$$

Thus in studying quadratic equations mod p, it suffices to consider equations of the form

$$x^2 \equiv a \mod p.$$

If $p|a$ we have the uninteresting equation $x^2 \equiv 0$, hence $x \equiv 0$, mod p. Thus assume that p does not divide a.

A2. Definition

The *Legendre symbol*

$$\chi(a) = \left(\frac{a}{p}\right)$$

is given by

$$\chi(a) = \begin{cases} 1 & \text{if } a^{(p-1)/2} \equiv 1 \mod p \\ -1 & \text{if } a^{(p-1)/2} \equiv -1 \mod p. \end{cases}$$

If $b = a^{(p-1)/2}$ then $b^2 = a^{p-1} \equiv 1 \mod p$, so $b \equiv \pm 1 \mod p$ and χ is well-defined. Thus

$$\chi(a) \equiv a^{(p-1)/2} \mod p.$$

A3. Theorem

The Legendre symbol $\left(\frac{a}{p}\right)$ is 1 if and only if a is a quadratic residue (from now on abbreviated QR) mod p.

Proof. If $a \equiv x^2 \mod p$ then $a^{(p-1)/2} \equiv x^{p-1} \equiv 1 \mod p$. (Note that if p divides x then p divides a, a contradiction.) Conversely, suppose $a^{(p-1)/2} \equiv 1 \mod p$. If g is a primitive root mod p, then $a \equiv g^r \mod p$ for some r. Therefore $a^{(p-1)/2} \equiv g^{r(p-1)/2} \equiv 1 \mod p$, so $p - 1$ divides $r(p-1)/2$, hence $r/2$ is an integer. But then $(g^{r/2})^2 = g^r \equiv a \mod p$, and a is a QR mod p. ♣

A4. Theorem

The mapping $a \to \chi(a)$ is a homomorphism from $\mathbb{F}_p$ to $\{\pm 1\}$.

Proof. We compute

$$\left(\frac{ab}{p}\right) \equiv (ab)^{(p-1)/2} = a^{(p-1)/2}b^{(p-1)/2} \equiv \left(\frac{a}{p}\right)\left(\frac{b}{p}\right)$$

so $\chi(ab) = \chi(a)\chi(b)$. ♣

A5. Theorem

If g is a primitive root mod p, then $\chi(g) = -1$, so g is a quadratic nonresidue (from now on abbreviated QNR) mod p. Consequently, exactly half of the integers $1,2,...,p-1$ are QR's and half are QNR's.

Proof. If $h^2 \equiv g \mod p$ then h is a primitive root with twice the period of g, which is impossible. Thus by (A4), $\chi(ag) = -\chi(a)$, so $a \to ag$ gives a bijection between QR's and QNR's. ♣

A6. The First Supplementary Law

From the definition (A2) and the fact that $(p-1)/2$ is even if $p \equiv 1 \mod 4$ and odd if $p \equiv 3 \mod 4$, we have

$$\left(\frac{-1}{p}\right) = (-1)^{(p-1)/2} = \begin{cases} 1 & \text{if } p \equiv 1 \mod 4 \\ -1 & \text{if } p \equiv 3 \mod 4. \end{cases}$$

A7. Definition

Let K be a field of characteristic $\neq p$ such that K contains the p-th roots of unity. Let $\zeta \in K$ be a primitive p-th root of unity. Define the *Gauss sum* by

$$\tau_p = \sum_{a=1}^{p-1} \left(\frac{a}{p}\right)\zeta^a.$$

A8. Theorem

$$\tau_p^2 = (-1)^{(p-1)/2} \, p.$$

Proof. From the definition of Gauss sum and (A4) we have

$$\tau_p^2 = \sum_{a,b=1}^{p-1} \left(\frac{ab}{p}\right)\zeta^{a+b}.$$

For each a, we can sum over all c such that $b \equiv ac \mod p$. (As c ranges over $1, \dots, p-1$, ac also takes all values $1, \dots, p-1$.) Thus

$$\tau_p^2 = \sum_{a=1}^{p-1} \sum_{c=1}^{p-1} \left(\frac{a^2 c}{p}\right) \zeta^{a+ac}.$$

Since $\left(\frac{a^2}{p}\right) = 1$, this simplifies to

$$\tau_p^2 = \sum_{c=1}^{p-1} \left(\sum_{a=1}^{p-1} \zeta^{a(1+c)}\right) \left(\frac{c}{p}\right).$$

If $1 + c \not\equiv 0 \mod p$ then $1, \zeta^{1+c}, \zeta^{2(1+c)}, \dots, \zeta^{(p-1)(1+c)}$ runs through all the roots (zeros) of $X^p - 1$ (note that $\zeta^p = 1$). But the coefficient of X^{p-1} is 0, so the sum of the roots is 0. Therefore the sum of $\zeta^{1+c}, \zeta^{2(1+c)}, \dots, \zeta^{(p-1)(1+c)}$ is -1. If $1 + c \equiv 0 \mod p$, then we are summing $p - 1$ ones. Consequently,

$$\tau_p^2 = -\sum_{c=1}^{p-2} \left(\frac{c}{p}\right) + (p-1)\left(\frac{-1}{p}\right).$$

(Note that $1 + c \equiv 0 \mod p \leftrightarrow c = p - 1$.) We can sum from 1 to $p - 1$ if we add $\left(\frac{-1}{p}\right)$, hence

$$\tau_p^2 = -\sum_{c=1}^{p-1} \left(\frac{c}{p}\right) + p\left(\frac{-1}{p}\right).$$

The sum is 0 by (A5), and the result follows from (A6). ♣

A9. The Law of Quadratic Reciprocity

Let p and q be odd primes, with $p \neq q$. Then

$$\left(\frac{p}{q}\right)\left(\frac{q}{p}\right) = (-1)^{(p-1)(q-1)/4}.$$

Thus if either p or q is congruent to 1 mod 4, then p is a QR mod q if and only if q is a QR mod p; and if both p and q are congruent to 3 mod 4, then p is a QR mod q if and only if q is a QNR mod p.

Proof. Let K have characteristic q and contain the p-th roots of unity. For example, take K to be the splitting field of $X^p - 1$ over $\mathbb{F}_q$. Then

$$\tau_p^q = (\tau_p^2)^{(q-1)/2} \tau_p$$

which by (A8) is

$$\left[(-1)^{(p-1)/2} \, p\right]^{(q-1)/2} \tau_p.$$

Thus

$$\tau_p^q = (-1)^{(p-1)(q-1)/4}\, p^{(q-1)/2}\, \tau_p.$$

But by the binomial expansion applied to the definition of τ_p in (A7),

$$\tau_p^q = \sum_{a=1}^{p-1} \left(\frac{a}{p}\right)\zeta^{aq}$$

(Recall that K has characteristic q, so $a^q = a$ in K.) Let $c \equiv aq \mod p$ and note that

$$\left(\frac{q^{-1}}{p}\right) = \left(\frac{q}{p}\right)$$

because the product of the two terms is $\left(\frac{1}{p}\right) = 1$. Thus

$$\tau_p^q = \sum_{c=1}^{p-1} \left(\frac{c}{p}\right)\left(\frac{q}{p}\right)\zeta^c = \left(\frac{q}{p}\right)\tau_p.$$

We now have two expressions (not involving summations) for τ_p^q, so

$$(-1)^{(p-1)(q-1)/4}\, p^{(q-1)/2} = \left(\frac{q}{p}\right).$$

Since $p^{(q-1)/2} = \left(\frac{p}{q}\right)$ by (A2), the above equation holds not only in K but in $\mathbb{F}_q$, hence can be written as a congruence mod q. Finally, multiply both sides by $\left(\frac{p}{q}\right)$ to complete the proof. ♣

A10. The Second Supplementary Law

$$\left(\frac{2}{p}\right) = (-1)^{(p^2-1)/8}$$

so

$$\left(\frac{2}{p}\right) = \begin{cases} 1 & \text{if } p \equiv \pm 1 \mod 8 \\ -1 & \text{if } p \equiv \pm 3 \mod 8. \end{cases}$$

Thus if $p \equiv 1$ or $7 \mod 8$, then 2 is a QR mod p, and if $p \equiv 3$ or $5 \mod 8$, then 2 is a QNR mod p.

Proof. Let K be a field of characteristic p containing the 8th roots of unity, and let ζ be a primitive 8th root of unity. Define $\tau = \zeta + \zeta^{-1}$. Then $\tau^2 = \zeta^2 + \zeta^{-2} + 2$. Now ζ^2 and ζ^{-2} are distinct 4th roots of unity, not ± 1, so they must be negatives of each other (analogous to i and $-i$ in $\mathbb{C}$). Therefore $\tau^2 = 2$. Modulo p we have

$$\tau^p = (\tau^2)^{(p-1)/2}\,\tau = 2^{(p-1)/2}\,\tau = \left(\frac{2}{p}\right)\tau = \left(\frac{2}{p}\right)(\zeta + \zeta^{-1}).$$

But by definition of τ, $\tau^p = (\zeta + \zeta^{-1})^p = \zeta^p + \zeta^{-p}$. Now (again as in C) $\zeta^p + \zeta^{-p}$ and $\zeta + \zeta^{-1}$ will coincide if $p \equiv \pm 1 \bmod 8$, and will be negatives of each other if $p \equiv \pm 3 \bmod 8$. In other words,

$$\zeta^p + \zeta^{-p} = (-1)^{(p^2-1)/8}(\zeta + \zeta^{-1}).$$

Equating the two expressions for τ^p, we get the desired result. We can justify the appeal to the complex plane by requiring that K satisfy the constraints $\zeta^2 + \zeta^{-2} = 0$ and $\zeta^p + \zeta^{-p} = (-1)^{(p^2-1)/8}(\zeta + \zeta^{-1})$. ♣

A11. Example

We determine whether 113 is a QR mod 127:

$\left(\frac{113}{127}\right) = \left(\frac{127}{113}\right)$ because $113 \equiv 1 \bmod 4$;

$\left(\frac{127}{113}\right) = \left(\frac{14}{113}\right)$ because $\left(\frac{a}{p}\right)$ depends only on the residue class of a mod p;

$\left(\frac{14}{113}\right) = \left(\frac{2}{113}\right)\left(\frac{7}{113}\right) = \left(\frac{7}{113}\right)$ by (A4) and the fact that $113 \equiv 1 \bmod 8$;

$\left(\frac{7}{113}\right) = \left(\frac{113}{7}\right)$ because $113 \equiv 1 \bmod 4$;

$\left(\frac{113}{7}\right) = \left(\frac{1}{7}\right)$ because $113 \equiv 1 \bmod 7$

and since $\left(\frac{1}{7}\right) = 1$ by inspection, 113 is a QR mod 127. Explicitly,

$113 + 13(127) = 1764 = 42^2$.

A12. The Jacobi Symbol

Let Q be an odd positive integer with prime factorization $Q = q_1 \cdots q_s$. The *Jacobi symbol* is defined by

$$\left(\frac{a}{Q}\right) = \prod_{i=1}^{s} \left(\frac{a}{q_i}\right).$$

It follows directly from the definition that

$$\left(\frac{a}{Q}\right)\left(\frac{a}{Q'}\right) = \left(\frac{a}{QQ'}\right), \quad \left(\frac{a}{Q}\right)\left(\frac{a'}{Q}\right) = \left(\frac{aa'}{Q}\right).$$

Also,

$$a \equiv a' \bmod Q \Rightarrow \left(\frac{a}{Q}\right) = \left(\frac{a'}{Q}\right)$$

because if $a \equiv a' \bmod Q$ then $a \equiv a' \bmod q_i$ for all $i = 1, \ldots, s$.

Quadratic reciprocity and the two supplementary laws can be extended to the Jacobi symbol, if we are careful.

A13. Theorem

If Q is an odd positive integer then

$$\left(\frac{-1}{Q}\right) = (-1)^{(Q-1)/2}.$$

Proof. We compute

$$\left(\frac{-1}{Q}\right) = \prod_{j=1}^{s}\left(\frac{-1}{q_j}\right) = \prod_{j=1}^{s}(-1)^{(q_j-1)/2} = (-1)^{\sum_{j=1}^{s}(q_j-1)/2}.$$

Now if a and b are odd then

$$\frac{ab-1}{2} - \left(\frac{a-1}{2} + \frac{b-1}{2}\right) = \frac{(a-1)(b-1)}{2} \equiv 0 \pmod 2,$$

hence

$$\frac{a-1}{2} + \frac{b-1}{2} \equiv \frac{ab-1}{2} \pmod 2.$$

We can apply this result repeatedly (starting with $a = q_1, b = q_2$) to get the desired formula. ♣

A14. Theorem

If Q is an odd positive integer, then

$$\left(\frac{2}{Q}\right) = (-1)^{(Q^2-1)/8}.$$

Proof. As in (A13),

$$\left(\frac{2}{Q}\right) = \prod_{j=1}^{s}\left(\frac{2}{q_j}\right) = \prod_{j=1}^{s}(-1)^{(q_j^2-1)/8}.$$

But if a and b are odd then

$$\frac{a^2b^2-1}{8} - \left(\frac{a^2-1}{8} + \frac{b^2-1}{8}\right) = \frac{(a^2-1)(b^2-1)}{8} \equiv 0 \pmod 8.$$

In fact $\frac{a^2-1}{8} = \frac{(a-1)(a+1)}{8}$ is an integer and $b^2 - 1 \equiv 0 \pmod 8$. (Just plug in $b = 1, 3, 5, 7$.) Thus

$$\frac{a^2-1}{8} + \frac{b^2-1}{8} \equiv \frac{a^2b^2-1}{8} \pmod 8$$

and we can apply this repeatedly to get the desired result. ♣

A15. Theorem

If P and Q are odd, relatively prime positive integers, then

$$\left(\frac{P}{Q}\right)\left(\frac{Q}{P}\right) = (-1)^{(P-1)(Q-1)/4}.$$

Proof. Let the prime factorizations of P and Q be $P = \prod_{i=1}^{r} p_i$ and $Q = \prod_{j=1}^{s} q_j$. Then

$$\left(\frac{P}{Q}\right)\left(\frac{Q}{P}\right) = \prod_{i,j}\left(\frac{p_i}{q_j}\right)\left(\frac{q_j}{p_i}\right) = (-1)^{\sum_{i,j}(p_i-1)(q_j-1)/4}.$$

But as in (A13),

$$\sum_{i=1}^{r}(p_i - 1)/2 \equiv (P - 1)/2 \quad \text{mod } 2, \qquad \sum_{j=1}^{s}(q_j - 1)/2 \equiv (Q - 1)/2 \quad \text{mod } 2.$$

Therefore

$$\left(\frac{P}{Q}\right)\left(\frac{Q}{P}\right) = (-1)^{[(P-1)/2][(Q-1)/2]}$$

as desired. ♣

A16. Remarks

Not every property of the Legendre symbol extends to the Jacobi symbol. for example, $\left(\frac{2}{15}\right) = \left(\frac{2}{3}\right)\left(\frac{2}{5}\right) = (-1)(-1) = 1$, but 2 is a QNR mod 15.

B. Extension of Absolute Values

B1. Theorem

Let L/K be a finite extension of fields, with $n = [L : K]$. If $|\ |$ is an absolute value on K and K is locally compact (hence complete) in the topology induced by $|\ |$, then there is exactly one extension of $|\ |$ to an absolute value on L, namely

$$|a| = |N_{L/K}(a)|^{1/n}.$$

We will need to do some preliminary work.

B2. Lemma

Suppose we are trying to prove that $|\ |$ is an absolute value on L. Assume that $|\ |$ satisfies the first two requirements in the definition of absolute value in (9.1.1). If we find a real number $C > 0$ such that for all $a \in L$,

$$|a| \leq 1 \Rightarrow |1 + a| \leq C.$$

Then $|\ |$ satisfies the triangle inequality ($|a + b| \leq |a| + |b|$).

Proof. If $|a_1| \geq |a_2|$, then $a = a_2/a_1$ satisfies $|a| \leq 1$. We can take $C = 2$ without loss of generality (because we can replace C by $C^c = 2$). Thus

$$|a_1 + a_2| \leq 2a_1 = 2 \max\{|a_1|, |a_2|\}$$

so by induction,

$$|a_1 + \cdots + a_{2^r}| \leq 2^r \max |a_j|.$$

If n is any positive integer, choose r so that $2^{r-1} \leq n \leq 2^r$. Then

$$|a_1 + \cdots + a_n| \leq 2^r \max |a_j| \leq 2n \max |a_j|.$$

(Note that $2^{r-1} \leq n \Rightarrow 2^r \leq 2n$. Also, we can essentially regard n as 2^r by introducing zeros.) Now

$$|a + b|^n = \left| \sum_{j=0}^{n} \binom{n}{j} a^j b^{n-j} \right| \leq 2(n+1) \max_j \left\{ \left| \binom{n}{j} \right| |a|^j |b|^{n-j} \right\}.$$

But $|m| = |1 + 1 + \cdots + 1| \leq 2m$ for $m \geq 1$, so

$$|a + b|^n \leq 4(n+1) \max_j \left\{ \binom{n}{j} |a|^j |b|^{n-j} \right\}.$$

The expression in braces is a single term in a binomial expansion, hence

$$|a + b|^n \leq 4(n+1)(|a| + |b|)^n.$$

Taking n-th roots, we have

$$|a + b| \leq [4(n+1)]^{1/n}(|a| + |b|).$$

The right hand side approaches ($|a| + |b|$) as $n \to \infty$ (take logarithms), and the result follows. ♣

B3. Uniqueness

Since L is a finite-dimensional vector space over K, any two extensions to L are equivalent as norms, and therefore induce the same topology. Thus (see Section 9.1, Problem 3) for some $c > 0$ we have $|\ |_1 = (|\ |_2)^c$. But $|a|_1 = |a|_2$ for every $a \in K$, so c must be 1.

B4. Proof of Theorem B1

By (B2), it suffices to find $C > 0$ such that $|a| \leq 1 \Rightarrow |1 + a| \leq C$. Let $b_1, \ldots, b_n$ be a basis for L over K. If $a = \sum_{i=1}^n c_i b_i$, then the *max norm* on L is defined by

$$|a|_0 = \max_{1 \leq i \leq n} |c_i|.$$

The topology induced by $|\ |_0$ is the product topology determined by n copies of K. With respect to this topology, $N_{L/K}$ is continuous (it is a polynomial). Thus $a \to |a|$ is a composition of continuous functions, hence continuous. Consequently, $|\ |$ is a nonzero continuous function on the compact set $S = \{a \in L : |a|_0 = 1\}$. So there exist $\delta, \Delta > 0$ such that for all $a \in S$ we have

$$o < \delta \leq |a| \leq \Delta.$$

If $0 \neq a \in L$ we can find $c \in K$ such that $|a|_0 = |c|$. (We have $a = \sum_{i=1}^n c_i b_i$, and if $|c_i|$ is the maximum of the $|c_j|, 1 \leq j \leq n$, take $c = c_i$.) Then $|a/c|_0 = 1$, so $a/c \in S$ and

$$0 < \delta \leq |a/c| = \frac{|a/c|}{|a/c|_0} \leq \Delta.$$

Now

$$\frac{|a/c|}{|a/c|_0} = \frac{|a|}{|a|_0}$$

because $|a/c|_0 = |a|_0/|c|$. Therefore

$$0 < \delta \leq \frac{|a|}{|a|_0} \leq \Delta.$$

(Now suppose $|a| \leq 1$, so $|a|_0 \leq |a|/\delta \leq \delta^{-1}$. Thus

$$
\begin{aligned}
|1 + a| &\leq \Delta |1 + a|_0 \\
&\leq \Delta[|1|_0 + |a|_0] \qquad (1) \\
&\leq \Delta[|1|_0 + \delta^{-1}] = C
\end{aligned}
$$

where step (1) follows because $|\ |_0$ is a norm. ♣

C. The Different

C1. Definition

Let $\mathcal{O}_K$ be the ring of algebraic integers in the number field K. Let $\omega_1, \dots, \omega_n$ be an integral basis for $\mathcal{O}_K$, so that the field discriminant d_K is $\det T(\omega_i \omega_j))$, where T stands for trace. Define

$$\mathcal{D}^{-1} = \{x \in K : T(x\mathcal{O}_K) \subseteq \mathbb{Z}\}.$$

C2. Theorem

$\mathcal{D}^{-1}$ is a fractional ideal with $\mathbb{Z}$-basis $\omega_1^*, \dots, \omega_n^*$, the dual basis of $\omega_1, \dots, \omega_n$ referred to the vector space K over $\mathbb{Q}$. [The dual basis is determined by $T(\omega_i \omega_j^*) = \delta_{ij}$, see (2.2.9).]

Proof. In view of (3.2.5), if we can show that $\mathcal{D}^{-1}$ is an $\mathcal{O}_K$-module, it will follow that $\mathcal{D}^{-1}$ is a fractional ideal. We have

$$x \in \mathcal{O}_K, y \in \mathcal{D}^{-1} \Rightarrow T(xy\mathcal{O}_K) \subseteq T(y\mathcal{O}_K) \subseteq \mathbb{Z}$$

so $xy \in \mathcal{D}^{-1}$.

By (2.2.2), the trace of $\omega_i^* \omega_j$ is an integer for all j. Thus $\mathbb{Z}\omega_1^* + \cdots + \mathbb{Z}\omega_n^* \subseteq \mathcal{D}^{-1}$. We must prove the reverse inclusion. Let $x \in \mathcal{D}^{-1}$, so $x = \sum_{i=1}^n a_i \omega_i^*, a_i \in \mathbb{Q}$. Then

$$T(x\omega_j) = T(\sum_{i=1}^n a_i \omega_i^* \omega_j) = a_j.$$

But $x \in \mathcal{D}^{-1}$ implies that $T(x\omega_j) \in \mathbb{Z}$, so $a_j \in \mathbb{Z}$ and

$$\mathcal{D}^{-1} = \sum_{i=1}^n \mathbb{Z}\omega_i^*. \quad \clubsuit$$

C3. Remarks

Since K is the fraction field of $\mathcal{O}_K$, for each i there exists $a_i \in \mathcal{O}_K$ such that $a_i \omega_i^* \in \mathcal{O}_K$. By (2.2.8), we can take each a_i to be an integer. If $m = \prod_{i=1}^n a_i$, then $m\mathcal{D}^{-1} \subseteq \mathcal{O}_K$, which gives another proof that $\mathcal{D}^{-1}$ is a fractional ideal.

C4. Definition and Discussion

The *different* of K, denoted by $\mathcal{D}$, is the fractional ideal that is inverse to $\mathcal{D}^{-1}$; $\mathcal{D}^{-1}$ is called the *co-different*. In fact, $\mathcal{D}$ is an integral ideal of $\mathcal{O}_K$. We have $1 \in \mathcal{D}^{-1}$ by definition of $\mathcal{D}^{-1}$ and (2.2.2). Thus

$$\mathcal{D} = \mathcal{D}1 \subseteq \mathcal{D}\mathcal{D}^{-1} = \mathcal{O}_K.$$

The different can be defined in the general *AKLB* setup if A is integrally closed, so (2.2.2) applies.

C5. Theorem

The norm of $\mathcal{D}$ is $N(\mathcal{D}) = |d_K|$.

Proof. Let m be a positive integer such that $m\mathcal{D}^{-1} \subseteq \mathcal{O}_K$ [(see (C3)]. We have

$$m\omega_i^* = \sum_{j=1}^n a_{ij}\omega_j$$

$$\omega_i = \sum_{j=1}^n b_{ij}\omega_j^*$$

so there is a matrix equation $(b_{ij}) = (a_{ij}/m)^{-1}$. Now

$$T(\omega_i\omega_j) = \sum_{k=1}^n b_{ik}T(\omega_k^*\omega_j) = \sum_{k=1}^n b_{ik}\delta_{kj} = b_{ij}$$

so

$$\det(b_{ij}) = d_K.$$

By (C2), $m\mathcal{D}^{-1}$ is an ideal of $\mathcal{O}_K$ with $\mathbb{Z}$-basis $m\omega_i^*, i = 1\ldots,n$, so by (4.2.5),

$$d_{K/\mathbb{Q}}(m\omega_1^*, \ldots, m\omega_n^*) = N(m\mathcal{D}^{-1})^2 d_K$$

and by (2.3.2), the left side of this equation is $(\det a_{ij})^2 d_K$. Thus

$$|\det(a_{ij})| = N(m\mathcal{D}^{-1}) = m^n N(\mathcal{D}^{-1})$$

where the last step follows because $|B/I| = |B/mI|/|I/mI|$. Now $\mathcal{D}\mathcal{D}^{-1} = \mathcal{O}_K$ implies that $N(\mathcal{D}^{-1}) = N(\mathcal{D})^{-1}$, so

$$|d_K| = |\det(b_{ij})| = |\det(a_{ij}/m)|^{-1} = [N(\mathcal{D}^{-1})]^{-1} = N(\mathcal{D}). \quad \clubsuit$$

C6. Some Computations

We calculate the different of $K = \mathbb{Q}(\sqrt{-2})$. By (C5), $N(\mathcal{D}_K) = |d_K| = 4 \times 2 = 8$. Now $\mathcal{O}_K = \mathbb{Z}[\sqrt{-2}]$ is a principal ideal domain (in fact a Euclidean domain) so $\mathcal{D} = (a+b\sqrt{-2})$ for some $a, b \in \mathbb{Z}$. Taking norms, we have $8 = a^2 + 2b^2$, and the only integer solution is $a = 0, b = \pm 2$. Thus

$$\mathcal{D} = (2\sqrt{-2}) = (-2\sqrt{-2}).$$

We calculate the different of $K = \mathbb{Q}(\sqrt{-3})$. By (C5), $N(\mathcal{D}_K) = |d_K| = 3$. Now

$$\mathcal{O}_K = \mathbb{Z}[\omega], \quad \omega = -\frac{1}{2} + \frac{1}{2}\sqrt{-3}$$

and since $\mathcal{O}_K$ is a PID, $\mathcal{D} = (a + b\omega)$ for some $a, b \in \mathbb{Z}$. Taking norms, we get

$$3 = \left(a - \frac{b}{2}\right)^2 + 3\left(\frac{b}{2}\right)^2, \quad (2a - b)^2 + 3b^2 = 12.$$

There are 6 integer solutions:

$$2 + \omega, \quad -1 + \omega, \quad -2 - \omega, \quad 1 - \omega, \quad 1 + 2\omega, \quad -1 - 2\omega$$

but all of these elements are associates, so they generate the same principal ideal. Thus $\mathcal{D} = (2 + \omega)$.

It can be shown that a prime p ramifies in the number field K if and only if p divides the different of K.

Solutions to Problems

Chapter 1

Section 1.1

1. Multiply the equation by a^{n-1} to get

$$a^{-1} = -(c_{n-1} + \cdots + c_1 a^{n-2} + c_0 a^{n-1}) \in A.$$

2. Since $A[b]$ is a subring of B, it is an integral domain. Thus if $bz = 0$ and $b \neq 0$, then $z = 0$.

3. Any linear transformation on a finite-dimensional vector space is injective iff it is surjective. Thus if $b \in B$ and $b \neq 0$, there is an element $c \in A[b] \subseteq B$ such that $bc = 1$. Therefore B is a field.

4. Since P is the preimage of Q under the inclusion map of A into B, P is a prime ideal. The map $a + P \to a + Q$ is a well-defined injection of A/P into B/Q, since $P = Q \cap A$. Thus A/P can be viewed as a subring of B/Q.

5. If $b + Q \in B/Q$, then b satisfies an equation of the form

$$x^n + a_{n-1}x^{n-1} + \cdots + a_1 x + a_0 = 0, a_i \in A.$$

By Problem 4, $b + Q$ satisfies the same equation with a_i replaced by $a_i + P$ for all i. Thus B/Q is integral over A/P.

6. By Problems 1-3, A/P is a field if and only if B/Q is a field, and the result follows. (Note that B/Q is an integral domain (because Q is a prime ideal), as required in the hypothesis of the result just quoted.)

Section 1.2

1. If $x \notin \mathcal{M}$, then by maximality of $\mathcal{M}$, the ideal generated by $\mathcal{M}$ and x is R. Thus there exists $y \in \mathcal{M}$ and $z \in R$ such that $y + zx = 1$. By hypothesis, zx, hence x, is a unit. Take the contrapositive to conclude that $\mathcal{M}$ contains all nonunits, so R is a local ring by (1.2.8).

2. Any additive subgroup of the cyclic additive group of $\mathbb{Z}/p^n\mathbb{Z}$ must consist of multiples of some power of p, and it follows that every proper ideal is contained in (p), which must therefore be the unique maximal ideal.

3. The set of nonunits is $\mathcal{M} = \{f/g : g(a) \neq 0, f(a) = 0\}$, which is an ideal. By (1.2.8),

R is a local ring with maximal ideal $\mathcal{M}$.

4. $S^{-1}(g \circ f)$ takes m/s to $g(f(m))/s$, as does $(S^{-1}g) \circ (S^{-1}f)$. If f is the identity on M, then $S^{-1}f$ is the identity on $S^{-1}M$.

5. By hypothesis, $g \circ f = 0$, so $(S^{-1}g) \circ (S^{-1}f) = S^{-1}(g \circ f) = S^{-1}0 = 0$. Thus im $S^{-1}f \subseteq$ ker $S^{-1}g$. Conversely, let $y \in N, s \in S$, with $y/s \in$ ker $S^{-1}g$. Then $g(y)/s = 0/1$, so for some $t \in S$ we have $tg(y) = g(ty) = 0$. Therefore $ty \in$ ker $g = $ im f, so $ty = f(x)$ for some $x \in M$. We now have $y/s = ty/st = f(x)/st = (S^{-1}f)(x/st) \in$ im $S^{-1}f$.

6. The sequence $0 \to N \to M \to M/N \to 0$ is exact, so by Problem 5, the sequence $0 \to N_S \to M_S \to (M/N)_S \to 0$ is exact. (If f is one of the maps of the first sequence, the corresponding map in the second sequence is $S^{-1}f$.) It follows from the definition of localization of a module that $N_S \le M_S$, and by exactness of the second sequence we have $(M/N)_S \cong M_S/N_S$.

Section 2.1

1. A basis for $E/\mathbb{Q}$ is $1, \theta, \theta^2$, and

$$\theta^2 1 = \theta^2, \quad \theta^2\theta = \theta^3 = 3\theta - 1, \quad \theta^2\theta^2 = \theta^4 = \theta\theta^3 = 3\theta^2 - \theta.$$

Thus

$$m(\theta^2) = \begin{bmatrix} 0 & -1 & 0 \\ 0 & 3 & -1 \\ 1 & 0 & 3 \end{bmatrix}$$

and we have $T(\theta^2) = 6$, $N(\theta^2) = 1$. Note that if we had already computed the norm of θ (the matrix of θ is

$$m(\theta) = \begin{bmatrix} 0 & 0 & -1 \\ 1 & 0 & 3 \\ 0 & 1 & 0 \end{bmatrix}$$

and $T(\theta) = 0$, $N(\theta) = -1$), it would be easier to calculate $N(\theta^2)$ as $[N(\theta)]^2 = (-1)^2 = 1$.

2. The cyclotomic polynomial Ψ_6 has only two roots, ω and its complex conjugate $\overline{\omega}$. By (2.1.5),

$$T(\omega) = \omega + \overline{\omega} = e^{i\pi/3} + e^{-i\pi/3} = 2\cos\pi/3 = 1.$$

3. We have $\min(\theta, \mathbb{Q}) = X^4 - 2$, $\min(\theta^2, \mathbb{Q}) = X^2 - 2$, $\min(\theta^3, \mathbb{Q}) = X^4 - 8$, and $\min(\sqrt{3}\theta, \mathbb{Q}) = X^4 - 18$. (To compute the last two minimal polynomials, note that $(\theta^3)^4 = (\theta^4)^3 = 2^3 = 8$ and $(\sqrt{3}\theta)^4 = 18$.) Therefore all four traces are 0.

4. Suppose that $\sqrt{3} = a + b\theta + c\theta^2 + d\theta^3$. Take the trace of both sides to conclude that $a = 0$. (The trace of $\sqrt{3}$ is 0 because its minimal polynomial is $X^2 - 3$.) Thus $\sqrt{3} = b\theta + c\theta^2 + d\theta^3$, so $\sqrt{3}\theta = b\theta^2 + c\theta^3 + 2d$. Again take the trace of both sides to get $d = 0$. We now have $\sqrt{3} = b\theta + c\theta^2$, so $\sqrt{3}\theta^2 = b\theta^3 + 2c$. The minimal polynomial of $\sqrt{3}\theta^2$ is $X^2 - 6$, because $(\sqrt{3}\theta^2)^2 = 6$. Once again taking the trace of both sides, we get $c = 0$. Finally, $\sqrt{3} = b\theta$ implies $9 = 2b^4$, and we reach a contradiction.

Section 2.2

1. By the quadratic formula, $L = \mathbb{Q}(\sqrt{b^2 - 4c})$. Since $b^2 - 4c \in \mathbb{Q}$, we may write $b^2 - 4c = s/t = st/t^2$ for relatively prime integers s and t. We also have $s = uy^2$ and $t = vz^2$, with u and v relatively prime and square-free. Thus $L = \mathbb{Q}(\sqrt{uv}) = \mathbb{Q}(\sqrt{m})$.

2. If $\mathbb{Q}(\sqrt{d}) = \mathbb{Q}(\sqrt{e})$, then $\sqrt{d} = a + b\sqrt{e}$ for rational numbers a and b. Squaring both sides, we have $d = a^2 + b^2e + 2ab\sqrt{e}$, so $\sqrt{e}$ is rational, a contradiction (unless $a = 0$ and $b = 1$).

3. Any isomorphism of $\mathbb{Q}(\sqrt{d})$ and $\mathbb{Q}(\sqrt{e})$ must carry $\sqrt{d}$ into $a+b\sqrt{e}$ for rational numbers a and b. Thus d is mapped to $a^2 + b^2 + 2ab\sqrt{e}$. But a $\mathbb{Q}$-isomorphism maps d to d, and we reach a contradiction as in Problem 2.

4. Since $\omega_n = \omega_{2n}^2$, we have $\omega_n \in \mathbb{Q}(\omega_{2n})$, so $\mathbb{Q}(\omega_n) \subseteq \mathbb{Q}(\omega_{2n})$. If n is odd, then $n+1 = 2r$, so

$$\omega_{2n} = -\omega_{2n}^{2r} = -(\omega_{2n}^2)^r = -\omega_n^r.$$

Therefore $\mathbb{Q}(\omega_{2n}) \subseteq \mathbb{Q}(\omega_n)$.

5. $\mathbb{Q}(\sqrt{-3}) = \mathbb{Q}(\omega)$ where $\omega = -\frac{1}{2} + \frac{1}{2}\sqrt{-3}$ is a primitive cube root of unity.

6. If $l(y) = 0$, then $(x, y) = 0$ for all x. Since the bilinear form is nondegenerate, we must have $y = 0$.

7. Since V and V^* have the same dimension, the map $y \to l(y)$ is surjective.

8. We have $(x_i, y_j) = l(y_j)(x_i) = f_j(x_i) = \delta_{ij}$. Since the $f_j = l(y_j)$ form a basis, so do the y_j.

9. Write $x_i = \sum_{k=1}^n a_{ik} y_k$, and take the inner product of both sides with x_j to conclude that $a_{ij} = (x_i, x_j)$.

Section 2.3

1. The first statement follows because multiplication of each element of a group G by a particular element $g \in G$ permutes the elements of G. We can work in a Galois extension of $\mathbb{Q}$ containing L, and each automorphism in the Galois group restricts to one of the σ_i on L. Thus $P + N$ and PN belong to the fixed field of the Galois group, which is $\mathbb{Q}$.

2. Since the x_j are algebraic integers, so are the $\sigma_i(x_j)$, as in the proof of (2.2.2). Thus P and N, hence $P + N$ and PN, are algebraic integers. By (2.2.4), $P + N$ and PN belong to $\mathbb{Z}$.

3. $D = (P - N)^2 = (P + N)^2 - 4PN \equiv (P + N)^2 \mod 4$. But any square is congruent to 0 or 1 mod 4, and the result follows.

4. We have $y_i = \sum_{j=1}^n a_{ij}x_j$ with $a_{ij} \in \mathbb{Z}$. By (2.3.2), $D(y) = (\det A)^2 D(x)$. Since $D(y)$ is square-free, $\det A = \pm 1$, so A has an inverse with coefficients in $\mathbb{Z}$. Thus $x = A^{-1}y$, as claimed.

5. Every algebraic integer can be expressed as a $\mathbb{Z}$-linear combination of the x_i, hence of the y_i by Problem 4. Since the y_i form a basis for L over $\mathbb{Q}$, they are linearly independent and the result follows.

6. No. For example, take $L = \mathbb{Q}(\sqrt{m})$, where m is a square-free integer with $m \not\equiv 1 \mod 4$. By (2.3.11), the field discriminant is $4m$, which is not square-free.

Section 3.1

1. We may assume that I is not contained in the union of any collection of $s - 1$ of the P_i's. (If so, we can simply replace s by $s - 1$.) It follows that elements of the desired form exist.

2. Assume that $I \not\subseteq P_1$ and $I \not\subseteq P_2$. We have $a_1 \in P_1$, $a_2 \notin P_1$, so $a_1 + a_2 \notin P_1$. Similarly, $a_1 \notin P_2$, $a_2 \in P_2$, so $a_1 + a_2 \notin P_2$. Thus $a_1 + a_2 \notin I \subseteq P_1 \cup P_2$, contradicting $a_1, a_2 \in I$.

3. For all $i = 1, \ldots, s - 1$ we have $a_i \notin P_s$, hence $a_1 \cdots a_{s-1} \notin P_s$ because P_s is prime. But $a_s \in P_s$, so a cannot be in P_s. Thus $a \in I$ and $a \notin P_1 \cup \cdots \cup P_s$.

Section 3.2

1. The product of ideals is always contained in the intersection. If I and J are relatively prime, then $1 = x + y$ with $x \in I$ and $y \in J$. If $z \in I \cap J$, then $z = z1 = zx + zy \in IJ$. The general result follows by induction, along with the computation

$$R = (I_1 + I_3)(I_2 + I_3) \subseteq I_1 I_2 + I_3.$$

Thus $I_1 I_2$ and I_3 are relatively prime. Continue in this manner with

$$R = (I_1 I_2 + I_4)(I_3 + I_4) \subseteq I_1 I_2 I_3 + I_4$$

and so on.

2. We have $R = R^r = (P_1 + P_2)^r \subseteq P_1^r + P_2$. Thus P_1^r and P_2 are relatively prime for all $r \geq 1$. Assuming inductively that P_1^r and P_2^s are relatively prime, it follows that

$$P_2^s = P_2^s R = P_2^s (P_1^r + P_2) \subseteq P_1^r + P_2^{s+1}$$

so

$$R = P_1^r + P_2^s \subseteq P_1^r + (P_1^r + P_2^{s+1}) = P_1^r + P_2^{s+1}$$

completing the induction.

3. Let r be a nonzero element of R such that $rK \subseteq R$, hence $K \subseteq r^{-1}R \subseteq K$. Thus $K = r^{-1}R$. Since $r^{-2} \in K$ we have $r^{-2} = r^{-1}s$ for some $s \in R$. But then $r^{-1} = s \in R$, so $K \subseteq R$ and consequently $K = R$.

Section 3.3

1. By (2.1.10), the norms are 6,6,4 and 9. Now if $x = a + b\sqrt{-5}$ and $x = yz$, then $N(x) = a^2 + 5b^2 = N(y)N(z)$. The only algebraic integers of norm 1 are ± 1, and there are no algebraic integers of norm 2 or 3. Thus there cannot be a nontrivial factorization of $1 \pm \sqrt{-5}$, 2 or 3.

2. If $(a + b\sqrt{-5})(c + d\sqrt{-5}) = 1$, take norms to get $(a^2 + 5b^2)(c^2 + 5d^2) = 1$, so $b = d = 0$, $a = \pm 1, c = \pm 1$.

3. By Problem 2, if two factors are associates, then the quotient of the factors is ± 1, which is impossible.

4. This is done as in Problems 1-3, using $18 = (1 + \sqrt{-17})(1 - \sqrt{-17}) = 2 \times 3^2$.

5. By (2.2.6) or (2.3.11), the algebraic integers are of the form $a + b\sqrt{-3}, a, b \in \mathbb{Z}$, or $(u/2) + (v/2)\sqrt{-3}$ with u and v odd integers. If we require that the norm be 1, we only get ± 1 in the first case. But in the second case, we have $u^2 + 3v^2 = 4$, so $u = \pm 1, v = \pm 1$. Thus if $\omega = e^{i2\pi/3}$, then the algebraic integers of norm 1 are ± 1, $\pm \omega$, and $\pm \omega^2$.

Section 3.4

1. $1 - \sqrt{-5} = 2 - (1 + \sqrt{-5}) \in P_2$, so $(1 + \sqrt{-5})(1 - \sqrt{-5}) = 6 \in P_2^2$.
2. Since $2 \in P_2$, it follows that $4 \in P_2^2$, so by Problem 1, $2 = 6 - 4 \in P_2^2$.
3. $(2, 1 + \sqrt{-5})(2, 1 + \sqrt{-5}) = (4, 2(1 + \sqrt{-5}), (1 + \sqrt{-5})^2)$, and $(1 + \sqrt{-5})^2 = -4 + 2\sqrt{-5}$. Therefore each of the generators of the ideal P_2^2 is divisible by 2, hence belongs to (2). Thus $P_2^2 \subseteq (2)$.
4. $x^2 + 5 \equiv (x+1)(x-1) \mod 3$, which suggests that $(3) = P_3 P_3'$, where $P_3 = (3, 1 + \sqrt{-5})$ and $P_3' = (3, 1 - \sqrt{-5})$.
5. $P_3 P_3' = (3, 3(1 + \sqrt{-5}), 3(1 - \sqrt{-5}), 6) \subseteq (3)$, because each generator of $P_3 P_3'$ is divisible by 3. But $3 \in P_3 \cap P_3'$, hence $9 \in P_3 P_3'$, and therefore $9 - 6 = 3 \in P_3 P_3'$. Thus $(3) \subseteq P_3 P_3'$, and the result follows.

Section 4.1

1. The kernel is $\{a \in A : a/1 \in \mathcal{M}S^{-1}A\} = A \cap (\mathcal{M}S^{-1}A) = \mathcal{M}$ by (1.2.6).
2. By hypothesis, $\mathcal{M} \cap S = \emptyset$, so $s \notin \mathcal{M}$. By maximality of $\mathcal{M}$ we have $\mathcal{M} + As = A$, so $y + bs = 1$ for some $y \in \mathcal{M}, b \in A$. Thus $bs \equiv 1 \mod \mathcal{M}$.
3. Since $1 - bs \in \mathcal{M}$, $(a/s) - ab = (a/s)(1 - bs) \in \mathcal{M}S^{-1}A$. Therefore $(a/s) + \mathcal{M}S^{-1}A = ab + \mathcal{M}S^{-1}A = h(ab)$.

Section 4.2

1. By the Chinese remainder theorem, $B/(p) \cong \prod_i B/P_i^{e_i}$. If p does not ramify, then $e_i = 1$ for all i, so $B/(p)$ is a product of fields, hence has no nonzero nilpotents. On the other hand, suppose that $e = e_i > 1$, with $P = P_i$. Choose $x \in P^{e-1} \setminus P^e$ and observe that $(x + P^e)^e = x^e + P^e$ is zero in B/P^e.
2. The minimal polynomial of a nilpotent element is a power of X, and the result follows from (2.1.5).
3. Let $\beta = \sum_{i=1}^n b_i \omega_i$ with $b_i \in \mathbb{Z}$. Then, with T denoting trace,

$$T(A(\beta \omega_j)) = T(\sum_{i=1}^n b_i A(\omega_i \omega_j)) = \sum_{i=1}^n b_i T(\omega_i \omega_j) \equiv 0 \mod p.$$

If $\beta \notin (p)$, then not all the b_i can be 0 mod p, so the determinant of the matrix $(T(\omega_i \omega_j))$, which is the discriminant d by (2.3.1), is 0 mod p. Therefore, p divides d.
4. This follows from the Chinese remainder theorem, as in Problem 1. The fields F_i all have characteristic p because p annihilates $B/(p)$.
5. The T_i are nondegenerate by separability, and $\sum_i T_i$ is nondegenerate by orthogonality, that is, $\pi_i(x)\pi_j(y) = 0$ for $i \neq j$.

6. Since $F_i/\mathbb{F}_p$ is a finite extension of a finite field, it is a Galois extension, so all embeddings are actually automorphisms. Thus for any $z \in F_i$, the endomorphism given by multiplication by z has trace $T_{F_i/\mathbb{F}_p}(z) = T_i(z)$. Since $B/(p)$ is, in particular, a direct sum of the F_i, the result follows.

Section 4.3

1. Factoring (2) is covered by case (c1) of (4.3.2), and we have $(2) = (2, 1 + \sqrt{-5})^2$. Factoring (3) is covered by case (a1), and $x^2 + 5 \equiv (x + 1)(x - 1) \mod 3$. Therefore $(3) = (3, 1 + \sqrt{-5})\,(3, 1 - \sqrt{-5})$.

2. We have $(5) = (5, \sqrt{-5})^2$, as in case (b). To factor (7), note that $x^2 + 5$ factors mod 7 as $(x + 3)\,(x - 3)$, so $(7) = (7, 3 + \sqrt{-5})\,(7, 3 - \sqrt{-5})$, as in case (a1). Since -5 is not a quadratic residue mod 11, we are in case (a2) and 11 remains prime.

3. Mod 5 we have $x^3 - 2 \equiv x^3 - 27 = x^3 - 3^3 = (x - 3)(x^2 + 3x + 9) = (x + 2)(x^2 + 3x - 1)$. Thus

$$(5) = (5, \alpha + 2)(5, \alpha^2 + 3\alpha - 1)$$

where $\alpha = \sqrt[3]{2}$.

Section 5.3

1. We have $r_2 = 1$ and $n = 2$, so the bound is $(4/\pi)(2/4)\sqrt{|d|} = (2/\pi)\sqrt{|d|}$. The discriminant may be calculated from (2.3.11). We have $d = 4m$ for $m = -1, -2$, and $d = m$ for $m = -3, -7$. The largest $|d|$ is 8, and the corresponding bound is $4\sqrt{2}/\pi$, which is about 1.80. Thus all the class numbers are 1.

2. We have $r_2 = 0$ and $n = 2$, so the bound is $\sqrt{|d|}/2$. We have $d = 4m$ for $m = 2, 3$, and $d = m$ for $m = 5, 13$. The largest $|d|$ is 13, and the corresponding bound is $\sqrt{13}/2$, which is about 1.803. Thus all the class numbers are 1.

3. The discriminant is -20 and the Minkowski bound is $2\sqrt{20}/\pi$, which is about 2.85. Since 2 ramifies [see (4.3.2), case (c1)], there is only one ideal of norm 2. Thus the class number is at most 2. But we know that $\mathbb{Q}(\sqrt{-5})$ is not a UFD, by the exercises for Section 3.3. Therefore the class number is 2.

4. The discriminant is 24 and the bound is $\sqrt{24}/2 = \sqrt{6}$, which is about 2.45. Since 2 ramifies [see (4.3.2), case (b)], the argument proceeds as in Problem 3. Note that $\mathbb{Q}(\sqrt{6})$ is not a UFD because $-2 = (2 + \sqrt{6})(2 - \sqrt{6})$. Note also that $2 + \sqrt{6}$ and $2 - \sqrt{6}$ are associates, because $(2 + \sqrt{6})/(2 - \sqrt{6}) = -5 - 2\sqrt{6}$, which is a unit $[(-5 - 2\sqrt{6})(-5 + 2\sqrt{6}) = 1]$.

5. The discriminant is 17 and the bound is $\sqrt{17}/2$, which is about 2.06. Since 2 splits $[(4.3.2),$ case (c2)], there are 2 ideals of norm 2. In fact these ideals are principal, as can be seen from the factorization $-2 = [(3 + \sqrt{17})/2]\,[(3 - \sqrt{17})/2]$. Thus every ideal class contains a principal ideal, so the ideal class group is trivial.

6. The discriminant is 56 and the bound is $\sqrt{56}/2 = \sqrt{14}$, which is about 3.74. Since 3 remains prime $[(4.3.2),$ case (a2)], there are no ideals of norm 3. (The norm of the principal ideal (3) is $3^2 = 9$.) Since 2 ramifies $[(4.3.2),$ case (b)], there is only one ideal of norm 2. This ideal is principal, as can be seen from the factorization $2 = (4 + \sqrt{14})(4 - \sqrt{14})$. As in Problem 5, the class number is 1.

7. This follows from the Minkowski bound (5.3.5) if we note that $N(I) \geq 1$ and $2r_2 \leq n$.

8. By a direct computation, we get a_2 and

$$\frac{a_{n+1}}{a_n} = \frac{\pi}{4} \frac{(n+1)^{2n+2}}{n^{2n}} \frac{1}{(n+1)^2} = \frac{\pi}{4}(1 + \frac{1}{n})^{2n}.$$

By the binomial theorem, $a_{n+1}/a_n = (\pi/4)(1 + 2 + \text{positive terms}) \geq 3\pi/4$. Thus

$$|d| \geq a_2 \frac{a_3}{a_2} \cdots \frac{a_n}{a_{n-1}} \geq \frac{\pi^2}{4}(3\pi/4)^{n-2},$$

and we can verify by canceling common factors that $(\pi^2/4)(3\pi/4)^{n-2} \geq (\pi/3)(3\pi/4)^{n-1}$.

9. By Problem 8,

$$\log|d| \geq \log\frac{\pi}{3} + (n-1)\log\frac{3\pi}{4} = \log\frac{\pi}{3} - \log\frac{3\pi}{4} + n\log\frac{3\pi}{4}$$

and the result follows.

10. This follows from the bound given in Problem 8.

Section 6.1

1. Since x, hence jx, as well as e_i, hence $\lfloor jb_i \rfloor e_i$, all belong to H, so does x_j. We have $x_j \in T$ because $jb_i - \lfloor jb_i \rfloor \in [0, 1)$.

2. We have $x = x_1 + \sum_{i=1}^{r}\lfloor b_i \rfloor e_i$ with $x_1 \in H \cap T$ and the $e_i \in H \cap \overline{T}$. Since $H \cap T$ is a finite set, there are only finitely many choices for x_1. Since there are only finitely many e_i, H is finitely generated.

3. There are only finitely many distinct x_j and infinitely many integers, so $x_j = x_k$ for some $j \neq k$. By linear independence of the e_i, we have $(j - k)b_i = \lfloor jb_i \rfloor - \lfloor kb_i \rfloor$ for all i, and the result follows.

4. By the previous problems, H is generated by a finite number of elements that are linear combinations of the e_i with rational coefficients. If d is a common denominator of these coefficients, then $d \neq 0$ and $dH \subseteq \sum_{i=1}^{r} \mathbb{Z}e_i$. Thus dH is a subgroup of a free abelian group of rank r, hence is free of rank at most r.

5. Since $dH \cong H$, H is free, and since $H \supseteq \sum_{i=1}^{r} \mathbb{Z}e_i$, the rank of H is at least r, and hence exactly r.

Section 6.3

1. $m = 2 \Rightarrow 2 \times 1^2 = 1^2 + 1$, so the fundamental unit u is $1 + \sqrt{2}$ and we stop at step $t = 1$.

$m = 3 \Rightarrow 3 \times 1^2 = 2^2 - 1$, so $u = 2 + \sqrt{3}$ and $t = 1$.

$m = 5 \equiv 1 \mod 4 \Rightarrow 5 \times 1^2 = 1^2 + 4$, so $u = \frac{1}{2}(1 + \sqrt{5})$ and $t = 1$.

$m = 6 \Rightarrow 6 \times 2^2 = 5^2 - 1$, so $u = 5 + 2\sqrt{6}$ and $t = 2$.

$m = 7 \Rightarrow 7 \times 3^2 = 8^2 - 1$, so $u = 8 + 3\sqrt{7}$ and $t = 3$.

$m = 10 \Rightarrow 10 \times 1^2 = 3^2 + 1$, so $u = 3 + \sqrt{10}$ and $t = 1$.

$m = 11 \Rightarrow 11 \times 3^2 = 10^2 - 1$, so $u = 10 + 3\sqrt{11}$ and $t = 3$.

$m = 13 \equiv 1 \mod 4 \Rightarrow 13 \times 1^2 = 3^2 + 4$, so $u = \frac{1}{2}(3 + \sqrt{13})$ and $t = 1$.

$m = 14 \Rightarrow 14 \times 4^2 = 15^2 - 1$, so $u = 15 + 4\sqrt{14}$ and $t = 4$.
$m = 15 \Rightarrow 15 \times 1^2 = 4^2 - 1$, so $u = 4 + \sqrt{15}$ and $t = 1$.
$m = 17 \equiv 1 \mod 4 \Rightarrow 17 \times 2^2 = 8^2 + 4$, so $u = \frac{1}{2}(8 + 2\sqrt{17}) = 4 + \sqrt{17}$ and $t = 2$.
2. Note that $a/2$ and $b/2$ are both integers, so $u \in B_0$.
3. With $u = \frac{1}{2}(a + b\sqrt{m})$, we compute

$$8u^3 = a(a^2 + 3b^2m) + b(3a^2 + b^2m)\sqrt{m}.$$

Now $a^2 - b^2m = \pm 4$, and if we add $4b^2m$ to both sides, we get
$a^2 + 3b^2m = 4b^2m \pm 4 = 4(b^2m \pm 1)$. Since $m \equiv 1 \mod 4$, m must be odd, and since b is
also odd, $b^2m \pm 1$ is even, so $4(b^2m \pm 1)$ is divisible by 8. Similarly,
$3a^2 + b^2m = 4a^2 - (a^2 - b^2m) = 4a^2 \pm 4$, which is also divisible by 8 because a is odd. It
follows that $u^3 \in B_0$.
4. If $u^2 \in B_0$, then u^2 is a positive unit in B_0, hence so is $(u^2)^{-1} = u^{-2}$. Therefore
$u = u^3u^{-2} \in B_0$. But a and b are odd, so $u \notin \mathbb{Z}[\sqrt{m}]$, a contradiction.
5. When $m = 5$, we have $u = \frac{1}{2}(1 + \sqrt{5})$, so $8u^3 = 1 + 3\sqrt{5} + (3 \times 5) + 5\sqrt{5}$. Thus
$u^3 = 2 + \sqrt{5}$. Also, $4u^2 = 6 + 2\sqrt{5}$, so $u^2 = (3 + \sqrt{5})/2$. When $m = 13$, we have
$u = \frac{1}{2}(3 + \sqrt{13})$, so $8u^3 = 27 + 27\sqrt{13} + (3 \times 3 \times 13) + 13\sqrt{13}$. Therefore $u^3 = 18 + 5\sqrt{13}$.
Also, $4u^2 = 22 + 6\sqrt{13} = (11 + 3\sqrt{13})/2$.

Note that the results for u^3 in Problem 5 are exactly what we would get by solving
$a^2 - mb^2 = \pm 1$. For $m = 5$ we have $5 \times 1^2 = 2^2 + 1$, so $a = 2, b = 1$. For $m = 13$ we have
$13 \times 5^2 = 18^2 + 1$, so $a = 18, b = 5$.

Section 7.1

1. The missing terms in the product defining the discriminant are either squares of
real numbers or occur as a complex number and its conjugate. Thus the missing terms
contribute a positive real number, which cannot change the overall sign.
2. Observe that $(c - \bar{c})^2$ is a negative real number, so each pair of complex embeddings
contributes a negative sign.
3. We have $2r_2 = [\mathbb{Q}(\zeta) : \mathbb{Q}] = \varphi(p^r) = p^{r-1}(p - 1)$, so the sign is $(-1)^s$, where, assuming
$p^r > 2$, $s = p^{r-1}(p - 1)/2$. To show that there are no real embeddings, note that if ζ
is mapped to -1, then $-\zeta$ is mapped to 1. But 1 is also mapped to 1, and (assuming a
nontrivial extension), we reach a contradiction.

Examination of the formula for s allows further simplification. If p is odd, the sign
will be positive if and only if $p \equiv 1 \mod 4$. If $p = 2$, the sign will be positive iff $r > 2$.

Section 8.1

1. If $\tau \in I(Q)$ and $x \in B$, then

$$\sigma\tau\sigma^{-1}(x) - x = \sigma(\tau\sigma^{-1}(x) - \sigma^{-1}(x)) \in \sigma(Q)$$

so $\sigma I(Q)\sigma^{-1} \subseteq I(\sigma(Q))$. Conversely, let $\tau \in I(\sigma(Q))$, $x \in B$. Then $\tau = \sigma(\sigma^{-1}\tau\sigma)\sigma^{-1}$,
so we must show that $\sigma^{-1}\tau\sigma \in I(Q)$, in other words, $\sigma^{-1}\tau\sigma(x) - x \in Q$. Now we have
$\tau\sigma(x) - \sigma(x) \in \sigma(Q)$, so $\tau\sigma(x) - \sigma(x) = \sigma(y)$ for some $y \in Q$. Thus $\sigma^{-1}\tau\sigma(x) - x = y \in Q$,

the desired result.

2. Since G is abelian, $\sigma D(Q)\sigma^{-1} = \sigma\sigma^{-1}D(Q) = D(Q)$, so by Problem 1 and (8.1.2), all the decomposition groups are the same. The decomposition groups depend only on P because P determines the unique factorization of PB into prime ideals of B. The analysis is the same for the inertia groups.

Section 9.1

1. This follows from (6.1.5) and the observation that a root of unity must have absolute value 1.

2. If the characteristic is $p \neq 0$, then there are only p integers, and the result follows from (9.1.7).

3. Assume the absolute values equivalent. By nontriviality, there is an element y with $|y|_1 > 1$. Take $a = \log|y|_2/\log|y|_1$. For every x there is a real number b such that $|x|_1 = |y|_1^b$. Find a sequence of rational numbers s/t converging to b from above. Then $|x|_1 = |y|_1^b < |y|_1^{s/t}$, so $|x^t/y^s|_1 < 1$. By hypothesis, $|x^t/y^s|_2 < 1$, so $|x|_2 < |y|_2^{s/t}$. Let $s/t \to b$ to get $|x|_2 \leq |y|_2^b$. But by taking a sequence of rationals converging to b from below, we get $|x|_2 \geq |y|_2^b$, hence $|x|_2 = |y|_2^b$. To summarize,

$$|x|_1 = |y|_1^b \Rightarrow |x|_2 = |y|_2^b.$$

Taking logarithms (if $x \neq 0$), we have $\log|x|_2/\log|x|_1 = a$, hence $|x|_1^a = |x|_2$.

Section 9.2

1. Let $a = \pm\prod p_i^{r_i}$, hence $\|a\|_\infty = \prod p_i^{r_i}$. If p is one of the p_i, then $\|a\|_p = p_i^{-r_i}$, and if p is not one of the p_i, then $\|a\|_p = 1$. Thus only finitely many terms of the product are unequal to 1, and the infinite prime cancels the effect of the finite primes. The result follows.

Section 9.3

1. For each $i = 1, \ldots, n$, choose $y_i, z_i \in k$ such that $|y_i|_i > 1$ and $|z_i|_i < 1$. This is possible by (9.3.2). Take $x_i = y_i$ if $i \leq r$, and $x_i = z_i$ if $i > r$. By (9.3.3), there is an element $a \in k$ such that $|a - x_i|_i < \epsilon$ for all i. (We will specify ϵ in a moment.) If $i \leq r$, then

$$|y_i|_i \leq |y_i - a|_i + |a|_i < \epsilon + |a|_i$$

so $|a|_i > |y_i|_i - \epsilon$, and we need $0 < \epsilon \leq |y_i|_i - 1$. On the other hand, if $i > r$, then

$$|a|_i \leq |a - z_i|_i + |z_i|_i < \epsilon + |z_i|_i$$

so we need $0 < \epsilon \leq 1 - |z_i|_i$. Since there are only finitely many conditions to be satisfied, a single ϵ can be chosen, and the result follows.

2. Since $|\ |_1$ is nontrivial, there is an element b such that $|b|_1 < 1$. Then we have

$$|a|_1 = 1 \Rightarrow |a^n b|_1 < 1 \Rightarrow |a^n b|_2 < 1 \Rightarrow |a|_2^n |b|_2 < 1 \Rightarrow |a|_2 < 1/|b|_2^{1/n}.$$

Let $n \to \infty$ to get $|a|_2 \leq 1$. Similarly, $|a^{-1}|_2 \leq 1$, so $|a|_2 = 1$. Finally,

$$|a|_1 > 1 \Rightarrow |a^{-1}|_1 < 1 \Rightarrow |a^{-1}|_2 < 1 \Rightarrow |a|_2 > 1.$$

Note also that the implication $|a|_1 \geq 1 \Rightarrow |a|_2 \geq 1$ is sufficient for equivalence (consider the cntrapositive).

Section 9.4

1. The condition stated is equivalent to $v(a/b) \geq 0$.
2. The product is $4 + 2p + 4p^2 + p^3 + p^4$. But $4 = 1 + 3 = 1 + p$ and $4p^2 = p^2 + 3p^2 = p^2 + p^3$. Thus we have $1 + 3p + p^2 + 2p^3 + p^4 = 1 + 2p^2 + 2p^3 + p^4$.
3. We have $-1 = (p-1) - p = (p-1) + [(p-1) - p]p = (p-1) + (p-1)p - p^2$. Continuing inductively, we get

$$-1 = (p-1) + (p-1)p + (p-1)p^2 + \cdots.$$

The result can also be obtained by multiplying by -1 on each side of the equation

$$1 = (1-p)(1 + p + p^2 + \cdots).$$

4. Since $n! = 1 \cdot 2 \cdots p \cdots 2p \cdots 3p \cdots$, it follows that if $rp \leq n < (r+1)p$, then $|n!| = 1/p^r$. Thus $|n!| \to 0$ as $n \to \infty$.
5. No. Although $|p^r| = 1/p^r \to 0$ as $r \to \infty$, all integers n such that $rp < n < (r+1)p$ have absolute value 1. Thus the sequence of absolute values $|n|$ cannot converge, hence the sequence itself cannot converge.
6. We have $|a_n| = |1/n| = p^{v(n)}$, where $v(n)$ is the highest power of p dividing n. Thus $p^{v(n)} \leq n$, so $v(n) \leq \log n / \log p$ and consequently $v(n)/n \to 0$. We can apply the root test to get $\limsup |a_n|^{1/n} = \lim p^{v(n)/n} = 1$. The radius of convergence is the reciprocal of the lim sup, namely 1. Thus the series converges for $|x| < 1$ and diverges for $|x| > 1$. The series also diverges at $|x| = 1$ because $|1/n|$ does not converge to 0.
7. Since $\lfloor n/p^i \rfloor \leq n/p^i$ and $\sum_{i=1}^{\infty} 1/p^i = (1/p)/(1 - 1/p) = 1/(p-1)$, the result follows.
8. By Problem 7,

$$v[(p^m)!] = \frac{p^m}{p} + \frac{p^m}{p^2} + \cdots + \frac{p^m}{p^m} = 1 + p + \cdots + p^{m-1} = \frac{p^m - 1}{p - 1}.$$

9. We have $1/|n!| = p^{v(n!)} \leq p^{n/(p-1)}$ by Problem 7. Thus $|a_n|^{1/n} \leq p^{1/(p-1)}$. Thus the radius of convergence is at least $p^{-1/(p-1)}$. Now let $|x| = p^{-1/(p-1)} = (1/p)^{v(x)}$, so $v(x) = 1/(p-1)$. Taking $n = p^m$, we have, using Problem 8,

$$v(x^n/n!) = v[x^{p^m}/(p^m)!] = p^m v(x) - v[(p^m)!] = \frac{p^m}{p-1} - \frac{p^m - 1}{p - 1} = \frac{1}{p - 1}.$$

Since $1/(p-1)$ is a constant independent of m, $x^n/n!$ does not converge to 0, so the series diverges.

Note that $0 < 1/(p-1) < 1$, and since v is a discrete valuation, there is no $x \in \mathbb{Q}_p$ such that $v(x) = 1/(p-1)$. Thus $|x| < p^{-1/(p-1)}$ is equivalent to $|x| < 1$. But the sharper bound is useful in situations where $\mathbb{Q}_p$ is embedded in a larger field that extends the p-adic absolute value.

Section 9.5

1. Take $F(X) = X^{p-1} - 1$, which has $p - 1$ distinct roots mod p. (The multiplicative group of nonzero elements of $\mathbb{Z}/p\mathbb{Z}$ is cyclic.) All roots are simple (because $\deg F = p-1$). By (9.5.3), the roots lift to distinct roots of unity in $\mathbb{Z}_p$.

2. Take $F(X) = X^2 - m$. Since p does not divide m and $p \neq 2$, F and its derivative are relatively prime, so there are no multiple roots. By (9.5.3), m is a square in $\mathbb{Z}_p$ iff m is a quadratic residue mod p.

3. Successively find $a_0, a_1, \ldots$, such that $(a_0 + a_1 p + a_2 p^2 + \cdots)^2 = m$ in $\mathbb{Z}_p$. If we take $p = 5$, $m = 6$, then the first four coefficients are $a_0 = 1, a_1 = 3, a_2 = 0, a_3 = 4$. There is a second solution, the negative of this one. When computing, don't forget the carry. For example, $(1 + 3 \times 5^1 + a_2 \times 5^2 + \cdots)^2 = 1 + 1 \times 5^1$ yields a term $6 \times 5^1 = 1 \times 5^1 + 1 \times 5^2$, so the equation for a_2 is $2a_2 + 10$ (not 9) $\equiv 0 \mod 5$, so $a_2 = 0$.

Index

A CATALOG OF SELECTED
DOVER BOOKS
IN SCIENCE AND MATHEMATICS

Mathematics

FUNCTIONAL ANALYSIS (Second Corrected Edition), George Bachman and Lawrence Narici. Excellent treatment of subject geared toward students with background in linear algebra, advanced calculus, physics and engineering. Text covers introduction to inner-product spaces, normed, metric spaces, and topological spaces; complete orthonormal sets, the Hahn-Banach Theorem and its consequences, and many other related subjects. 1966 ed. 544pp. 6¹/₈ x 9¹/₄. 0-486-40251-7

DIFFERENTIAL MANIFOLDS, Antoni A. Kosinski. Introductory text for advanced undergraduates and graduate students presents systematic study of the topological structure of smooth manifolds, starting with elements of theory and concluding with method of surgery. 1993 edition. 288pp. 5³/₈ x 8¹/₂. 0-486-46244-7

VECTOR AND TENSOR ANALYSIS WITH APPLICATIONS, A. I. Borisenko and I. E. Tarapov. Concise introduction. Worked-out problems, solutions, exercises. 257pp. 5⅝ x 8¹/₄. 0-486-63833-2

AN INTRODUCTION TO ORDINARY DIFFERENTIAL EQUATIONS, Earl A. Coddington. A thorough and systematic first course in elementary differential equations for undergraduates in mathematics and science, with many exercises and problems (with answers). Index. 304pp. 5³/₈ x 8¹/₂. 0-486-65942-9

FOURIER SERIES AND ORTHOGONAL FUNCTIONS, Harry F. Davis. An incisive text combining theory and practical example to introduce Fourier series, orthogonal functions and applications of the Fourier method to boundary-value problems. 570 exercises. Answers and notes. 416pp. 5³/₈ x 8¹/₂. 0-486-65973-9

COMPUTABILITY AND UNSOLVABILITY, Martin Davis. Classic graduate-level introduction to theory of computability, usually referred to as theory of recurrent functions. New preface and appendix. 288pp. 5³/₈ x 8¹/₂. 0-486-61471-9

AN INTRODUCTION TO MATHEMATICAL ANALYSIS, Robert A. Rankin. Dealing chiefly with functions of a single real variable, this text by a distinguished educator introduces limits, continuity, differentiability, integration, convergence of infinite series, double series, and infinite products. 1963 edition. 624pp. 5³/₈ x 8¹/₂. 0-486-46251-X

METHODS OF NUMERICAL INTEGRATION (SECOND EDITION), Philip J. Davis and Philip Rabinowitz. Requiring only a background in calculus, this text covers approximate integration over finite and infinite intervals, error analysis, approximate integration in two or more dimensions, and automatic integration. 1984 edition. 624pp. 5³/₈ x 8¹/₂. 0-486-45339-1

INTRODUCTION TO LINEAR ALGEBRA AND DIFFERENTIAL EQUATIONS, John W. Dettman. Excellent text covers complex numbers, determinants, orthonormal bases, Laplace transforms, much more. Exercises with solutions. Undergraduate level. 416pp. 5³/₈ x 8¹/₂. 0-486-65191-6

RIEMANN'S ZETA FUNCTION, H. M. Edwards. Superb, high-level study of landmark 1859 publication entitled "On the Number of Primes Less Than a Given Magnitude" traces developments in mathematical theory that it inspired. xiv+315pp. 5³/₈ x 8¹/₂. 0-486-41740-9

CALCULUS OF VARIATIONS WITH APPLICATIONS, George M. Ewing. Applications-oriented introduction to variational theory develops insight and promotes understanding of specialized books, research papers. Suitable for advanced undergraduate/graduate students as primary, supplementary text. 352pp. 5³⁄₈ x 8¹⁄₂. 0-486-64856-7

MATHEMATICIAN'S DELIGHT, W. W. Sawyer. "Recommended with confidence" by *The Times Literary Supplement,* this lively survey was written by a renowned teacher. It starts with arithmetic and algebra, gradually proceeding to trigonometry and calculus. 1943 edition. 240pp. 5³⁄₈ x 8¹⁄₂. 0-486-46240-4

ADVANCED EUCLIDEAN GEOMETRY, Roger A. Johnson. This classic text explores the geometry of the triangle and the circle, concentrating on extensions of Euclidean theory, and examining in detail many relatively recent theorems. 1929 edition. 336pp. 5³⁄₈ x 8¹⁄₂. 0-486-46237-4

COUNTEREXAMPLES IN ANALYSIS, Bernard R. Gelbaum and John M. H. Olmsted. These counterexamples deal mostly with the part of analysis known as "real variables." The first half covers the real number system, and the second half encompasses higher dimensions. 1962 edition. xxiv+198pp. 5³⁄₈ x 8¹⁄₂. 0-486-42875-3

CATASTROPHE THEORY FOR SCIENTISTS AND ENGINEERS, Robert Gilmore. Advanced-level treatment describes mathematics of theory grounded in the work of Poincaré, R. Thom, other mathematicians. Also important applications to problems in mathematics, physics, chemistry and engineering. 1981 edition. References. 28 tables. 397 black-and-white illustrations. xvii + 666pp. 6¹⁄₈ x 9¹⁄₄. 0-486-67539-4

COMPLEX VARIABLES: Second Edition, Robert B. Ash and W. P. Novinger. Suitable for advanced undergraduates and graduate students, this newly revised treatment covers Cauchy theorem and its applications, analytic functions, and the prime number theorem. Numerous problems and solutions. 2004 edition. 224pp. 6¹⁄₂ x 9¹⁄₄. 0-486-46250-1

NUMERICAL METHODS FOR SCIENTISTS AND ENGINEERS, Richard Hamming. Classic text stresses frequency approach in coverage of algorithms, polynomial approximation, Fourier approximation, exponential approximation, other topics. Revised and enlarged 2nd edition. 721pp. 5³⁄₈ x 8¹⁄₂. 0-486-65241-6

INTRODUCTION TO NUMERICAL ANALYSIS (2nd Edition), F. B. Hildebrand. Classic, fundamental treatment covers computation, approximation, interpolation, numerical differentiation and integration, other topics. 150 new problems. 669pp. 5³⁄₈ x 8¹⁄₂. 0-486-65363-3

MARKOV PROCESSES AND POTENTIAL THEORY, Robert M. Blumental and Ronald K. Getoor. This graduate-level text explores the relationship between Markov processes and potential theory in terms of excessive functions, multiplicative functionals and subprocesses, additive functionals and their potentials, and dual processes. 1968 edition. 320pp. 5³⁄₈ x 8¹⁄₂. 0-486-46263-3

ABSTRACT SETS AND FINITE ORDINALS: An Introduction to the Study of Set Theory, G. B. Keene. This text unites logical and philosophical aspects of set theory in a manner intelligible to mathematicians without training in formal logic and to logicians without a mathematical background. 1961 edition. 112pp. 5³⁄₈ x 8¹⁄₂. 0-486-46249-8

INTRODUCTORY REAL ANALYSIS, A.N. Kolmogorov, S. V. Fomin. Translated by Richard A. Silverman. Self-contained, evenly paced introduction to real and functional analysis. Some 350 problems. 403pp. 5³/₈ x 8¹/₂.   0-486-61226-0

APPLIED ANALYSIS, Cornelius Lanczos. Classic work on analysis and design of finite processes for approximating solution of analytical problems. Algebraic equations, matrices, harmonic analysis, quadrature methods, much more. 559pp. 5³/₈ x 8¹/₂.   0-486-65656-X

AN INTRODUCTION TO ALGEBRAIC STRUCTURES, Joseph Landin. Superb self-contained text covers "abstract algebra": sets and numbers, theory of groups, theory of rings, much more. Numerous well-chosen examples, exercises. 247pp. 5³/₈ x 8¹/₂.
0-486-65940-2

QUALITATIVE THEORY OF DIFFERENTIAL EQUATIONS, V. V. Nemytskii and V.V. Stepanov. Classic graduate-level text by two prominent Soviet mathematicians covers classical differential equations as well as topological dynamics and ergodic theory. Bibliographies. 523pp. 5³/₈ x 8¹/₂.   0-486-65954-2

THEORY OF MATRICES, Sam Perlis. Outstanding text covering rank, nonsingularity and inverses in connection with the development of canonical matrices under the relation of equivalence, and without the intervention of determinants. Includes exercises. 237pp. 5³/₈ x 8¹/₂.   0-486-66810-X

INTRODUCTION TO ANALYSIS, Maxwell Rosenlicht. Unusually clear, accessible coverage of set theory, real number system, metric spaces, continuous functions, Riemann integration, multiple integrals, more. Wide range of problems. Undergraduate level. Bibliography. 254pp. 5³/₈ x 8¹/₂.   0-486-65038-3

MODERN NONLINEAR EQUATIONS, Thomas L. Saaty. Emphasizes practical solution of problems; covers seven types of equations. ". . . a welcome contribution to the existing literature. . . ."—*Math Reviews.* 490pp. 5³/₈ x 8¹/₂.   0-486-64232-1

MATRICES AND LINEAR ALGEBRA, Hans Schneider and George Phillip Barker. Basic textbook covers theory of matrices and its applications to systems of linear equations and related topics such as determinants, eigenvalues and differential equations. Numerous exercises. 432pp. 5³/₈ x 8¹/₂.   0-486-66014-1

LINEAR ALGEBRA, Georgi E. Shilov. Determinants, linear spaces, matrix algebras, similar topics. For advanced undergraduates, graduates. Silverman translation. 387pp. 5³/₈ x 8¹/₂.   0-486-63518-X

MATHEMATICAL METHODS OF GAME AND ECONOMIC THEORY: Revised Edition, Jean-Pierre Aubin. This text begins with optimization theory and convex analysis, followed by topics in game theory and mathematical economics, and concluding with an introduction to nonlinear analysis and control theory. 1982 edition. 656pp. 6¹/₈ x 9¹/₄.
0-486-46265-X

SET THEORY AND LOGIC, Robert R. Stoll. Lucid introduction to unified theory of mathematical concepts. Set theory and logic seen as tools for conceptual understanding of real number system. 496pp. 5⁵/₈ x 8¹/₄.   0-486-63829-4

Math—Geometry and Topology

ELEMENTARY CONCEPTS OF TOPOLOGY, Paul Alexandroff. Elegant, intuitive approach to topology from set-theoretic topology to Betti groups; how concepts of topology are useful in math and physics. 25 figures. 57pp. $5^3/_8$ x $8^1/_2$. 0-486-60747-X

A LONG WAY FROM EUCLID, Constance Reid. Lively guide by a prominent historian focuses on the role of Euclid's Elements in subsequent mathematical developments. Elementary algebra and plane geometry are sole prerequisites. 80 drawings. 1963 edition. 304pp. $5^3/_8$ x $8^1/_2$. 0-486-43613-6

EXPERIMENTS IN TOPOLOGY, Stephen Barr. Classic, lively explanation of one of the byways of mathematics. Klein bottles, Moebius strips, projective planes, map coloring, problem of the Koenigsberg bridges, much more, described with clarity and wit. 43 figures. 210pp. $5^3/_8$ x $8^1/_2$. 0-486-25933-1

THE GEOMETRY OF RENÉ DESCARTES, René Descartes. The great work founded analytical geometry. Original French text, Descartes's own diagrams, together with definitive Smith-Latham translation. 244pp. $5^3/_8$ x $8^1/_2$. 0-486-60068-8

EUCLIDEAN GEOMETRY AND TRANSFORMATIONS, Clayton W. Dodge. This introduction to Euclidean geometry emphasizes transformations, particularly isometries and similarities. Suitable for undergraduate courses, it includes numerous examples, many with detailed answers. 1972 ed. viii+296pp. $6^1/_8$ x $9^1/_4$. 0-486-43476-1

EXCURSIONS IN GEOMETRY, C. Stanley Ogilvy. A straightedge, compass, and a little thought are all that's needed to discover the intellectual excitement of geometry. Harmonic division and Apollonian circles, inversive geometry, hexlet, Golden Section, more. 132 illustrations. 192pp. $5^3/_8$ x $8^1/_2$. 0-486-26530-7

THE THIRTEEN BOOKS OF EUCLID'S ELEMENTS, translated with introduction and commentary by Sir Thomas L. Heath. Definitive edition. Textual and linguistic notes, mathematical analysis. 2,500 years of critical commentary. Unabridged. 1,414pp. $5^3/_8$ x $8^1/_2$. Three-vol. set.
Vol. I: 0-486-60088-2   Vol. II: 0-486-60089-0   Vol. III: 0-486-60090-4

SPACE AND GEOMETRY: IN THE LIGHT OF PHYSIOLOGICAL, PSYCHOLOGICAL AND PHYSICAL INQUIRY, Ernst Mach. Three essays by an eminent philosopher and scientist explore the nature, origin, and development of our concepts of space, with a distinctness and precision suitable for undergraduate students and other readers. 1906 ed. vi+148pp. $5^3/_8$ x $8^1/_2$. 0-486-43909-7

GEOMETRY OF COMPLEX NUMBERS, Hans Schwerdtfeger. Illuminating, widely praised book on analytic geometry of circles, the Moebius transformation, and two-dimensional non-Euclidean geometries. 200pp. $5^5/_8$ x $8^1/_4$. 0-486-63830-8

DIFFERENTIAL GEOMETRY, Heinrich W. Guggenheimer. Local differential geometry as an application of advanced calculus and linear algebra. Curvature, transformation groups, surfaces, more. Exercises. 62 figures. 378pp. $5^3/_8$ x $8^1/_2$. 0-486-63433-7

History of Math

THE WORKS OF ARCHIMEDES, Archimedes (T. L. Heath, ed.). Topics include the famous problems of the ratio of the areas of a cylinder and an inscribed sphere; the measurement of a circle; the properties of conoids, spheroids, and spirals; and the quadrature of the parabola. Informative introduction. clxxxvi+326pp. $5\frac{3}{8}$ x $8\frac{1}{2}$.   0-486-42084-1

A SHORT ACCOUNT OF THE HISTORY OF MATHEMATICS, W. W. Rouse Ball. One of clearest, most authoritative surveys from the Egyptians and Phoenicians through 19th-century figures such as Grassman, Galois, Riemann. Fourth edition. 522pp. $5\frac{3}{8}$ x $8\frac{1}{2}$.   0-486-20630-0

THE HISTORY OF THE CALCULUS AND ITS CONCEPTUAL DEVELOP-MENT, Carl B. Boyer. Origins in antiquity, medieval contributions, work of Newton, Leibniz, rigorous formulation. Treatment is verbal. 346pp. $5\frac{3}{8}$ x $8\frac{1}{2}$.   0-486-60509-4

THE HISTORICAL ROOTS OF ELEMENTARY MATHEMATICS, Lucas N. H. Bunt, Phillip S. Jones, and Jack D. Bedient. Fundamental underpinnings of modern arithmetic, algebra, geometry and number systems derived from ancient civilizations. 320pp. $5\frac{3}{8}$ x $8\frac{1}{2}$.   0-486-25563-8

THE HISTORY OF THE CALCULUS AND ITS CONCEPTUAL DEVELOP-MENT, Carl B. Boyer. Fluent description of the development of both the integral and differential calculus—its early beginnings in antiquity, medieval contributions, and a consideration of Newton and Leibniz. 368pp. $5\frac{3}{8}$ x $8\frac{1}{2}$.   0-486-60509-4

GAMES, GODS & GAMBLING: A HISTORY OF PROBABILITY AND STATISTICAL IDEAS, F. N. David. Episodes from the lives of Galileo, Fermat, Pascal, and others illustrate this fascinating account of the roots of mathematics. Features thought-provoking references to classics, archaeology, biography, poetry. 1962 edition. 304pp. $5\frac{3}{8}$ x $8\frac{1}{2}$. (Available in U.S. only.)   0-486-40023-9

OF MEN AND NUMBERS: THE STORY OF THE GREAT MATHEMATICIANS, Jane Muir. Fascinating accounts of the lives and accomplishments of history's greatest mathematical minds—Pythagoras, Descartes, Euler, Pascal, Cantor, many more. Anecdotal, illuminating. 30 diagrams. Bibliography. 256pp. $5\frac{3}{8}$ x $8\frac{1}{2}$.   0-486-28973-7

HISTORY OF MATHEMATICS, David E. Smith. Nontechnical survey from ancient Greece and Orient to late 19th century; evolution of arithmetic, geometry, trigonometry, calculating devices, algebra, the calculus. 362 illustrations. 1,355pp. $5\frac{3}{8}$ x $8\frac{1}{2}$. Two-vol. set.   Vol. I: 0-486-20429-4   Vol. II: 0-486-20430-8

A CONCISE HISTORY OF MATHEMATICS, Dirk J. Struik. The best brief history of mathematics. Stresses origins and covers every major figure from ancient Near East to 19th century. 41 illustrations. 195pp. $5\frac{3}{8}$ x $8\frac{1}{2}$.   0-486-60255-9

A TREATISE ON ELECTRICITY AND MAGNETISM, James Clerk Maxwell. Important foundation work of modern physics. Brings to final form Maxwell's theory of electromagnetism and rigorously derives his general equations of field theory. 1,084pp. 5³/₈ x 8¹/₂. Two-vol. set.   Vol. I: 0-486-60636-8   Vol. II: 0-486-60637-6

MATHEMATICS FOR PHYSICISTS, Philippe Dennery and Andre Krzywicki. Superb text provides math needed to understand today's more advanced topics in physics and engineering. Theory of functions of a complex variable, linear vector spaces, much more. Problems. 1967 edition. 400pp. 6¹/₂ x 9¹/₄.   0-486-69193-4

INTRODUCTION TO QUANTUM MECHANICS WITH APPLICATIONS TO CHEMISTRY, Linus Pauling & E. Bright Wilson, Jr. Classic undergraduate text by Nobel Prize winner applies quantum mechanics to chemical and physical problems. Numerous tables and figures enhance the text. Chapter bibliographies. Appendices. Index. 468pp. 5³/₈ x 8¹/₂.   0-486-64871-0

METHODS OF THERMODYNAMICS, Howard Reiss. Outstanding text focuses on physical technique of thermodynamics, typical problem areas of understanding, and significance and use of thermodynamic potential. 1965 edition. 238pp. 5³/₈ x 8¹/₂.
0-486-69445-3

THE ELECTROMAGNETIC FIELD, Albert Shadowitz. Comprehensive under- graduate text covers basics of electric and magnetic fields, builds up to electromagnetic theory. Also related topics, including relativity. Over 900 problems. 768pp. 5⁵/₈ x 8¹/₄.
0-486-65660-8

GREAT EXPERIMENTS IN PHYSICS: FIRSTHAND ACCOUNTS FROM GALILEO TO EINSTEIN, Morris H. Shamos (ed.). 25 crucial discoveries: Newton's laws of motion, Chadwick's study of the neutron, Hertz on electromagnetic waves, more. Original accounts clearly annotated. 370pp. 5³/₈ x 8¹/₂.   0-486-25346-5

EINSTEIN'S LEGACY, Julian Schwinger. A Nobel Laureate relates fascinating story of Einstein and development of relativity theory in well-illustrated, nontechnical volume. Subjects include meaning of time, paradoxes of space travel, gravity and its effect on light, non-Euclidean geometry and curving of space-time, impact of radio astronomy and space-age discoveries, and more. 189 b/w illustrations. xiv+250pp. 8³/₈ x 9¹/₄.   0-486-41974-6

THE VARIATIONAL PRINCIPLES OF MECHANICS, Cornelius Lanczos. Philosophic, less formalistic approach to analytical mechanics offers model of clear, scholarly exposition at graduate level with coverage of basics, calculus of variations, principle of virtual work, equations of motion, more. 418pp. 5³/₈ x 8¹/₂.   0-486-65067-7

Paperbound unless otherwise indicated. Available at your book dealer, online at www.doverpublications.com, or by writing to Dept. GI, Dover Publications, Inc., 31 East 2nd Street, Mineola, NY 11501. For current price information or for free catalogues (please indicate field of interest), write to Dover Publications or log on to www.doverpublications.com and see every Dover book in print. Dover publishes more than 400 books each year on science, elementary and advanced mathematics, biology, music, art, literary history, social sciences, and other areas.